TestSMART

for
Math, Grade 7

Book II, Operations & Problem Solving
Aligned to State and National Standards

Help for

Basic Math Skills

State Competency Tests

Achievement Tests

by

Lori Mammen

ecs

Copyright infringement is a violation of Federal Law.

© 2004 by ECS Learning Systems, Inc., San Antonio, Texas. All rights reserved. No part of this publication may be reproduced, translated, stored in a retrieval system, or transmitted in any way or by any means (electronic, mechanical, photocopying, recording, or otherwise) without prior written permission from ECS Learning Systems, Inc.

Photocopying of student worksheets by a classroom teacher at a non-profit school who has purchased this publication for his/her own class is permissible. Reproduction of any part of this publication for an entire school or for a school system, by for-profit institutions and tutoring centers, or for commercial sale is strictly prohibited.

TestSMART is a registered trademark of ECS Learning Systems, Inc. Printed in the United States of America.

NEW!

High Quality • Cost-Effective • Comprehensive

TestSMART® PLUS Software!

NEW Sequential Skill-building
Lessons & Practice Tests for State and National Standards

CAI (Computer Aided Instruction)	CAT (Computer Aided Testing)
Reading & Language Arts Grades 1–10 More than 1,935 lessons available!	**Reading & Language Arts** Grades K–8 More than 56 tests available!
Mathematics Grades K–10 More than 1,590 lessons available!	**Mathematics** Grades K–8 More than 48 tests available!

Updated to the latest National Standards! You receive MORE lessons and tests
with correlation to National Testing Standards for math, reading, and language arts!!
Single and multi-user licenses available

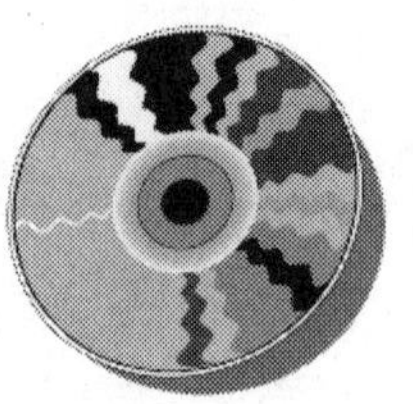

To order, or for more information,
visit www.educyberstor.com
or contact
ECS Learning Systems, Inc.
P.O. Box 791439, San Antonio, TX 78279.

Book Writers/Editors: Lori Mammen, Suzanne K. Mammen, Monica L. Odle **Book Cover:** Kirstin Simpson
Book Design: Educational Media Services **Page Layout & Graphics:** Jayme Salinas

ISBN 1-57022-249-5

Copyright infringement is a violation of Federal Law.

© 2004 by ECS Learning Systems, Inc., San Antonio, Texas. All rights reserved. No part of this publication may be reproduced, translated, stored in a retrieval system, or transmitted in any way or by any means (electronic, mechanical, photocopying, recording, or otherwise) without prior written permission from ECS Learning Systems, Inc.

Photocopying of student worksheets by a classroom teacher at a non-profit school who has purchased this publication for his/her own class is permissible. Reproduction of any part of this publication for an entire school or for a school system, by for-profit institutions and tutoring centers, or for commercial sale is strictly prohibited.

TestSMART is a registered trademark of ECS Learning Systems, Inc. Printed in the United States of America.

Contents

Notes

©ECS Learning Systems, Inc. All rights reserved

Welcome to *TestSMART*®!!

It's just the tool you need to help students review important mathematics skills and prepare for standardized mathematics tests!

Introduction

During the past several years, an increasing number of American students have faced some form of state-mandated competency testing in mathematics. While several states use established achievement tests, such as the Iowa Test of Basic Skills (ITBS), to assess students' achievement in mathematics, other states' assessments focus on the skills and knowledge emphasized in their particular mathematics curriculum. Texas, for example, has administered a state-developed assessment since 1980. The New York State Testing Program began in 1999 and tests both fourth- and eighth-grade students in mathematics.

Whatever the testing route, one point is very clear: the trend toward more and more competency testing is widespread and intense. By the spring of 1999, 48 states had adopted some type of assessment for students at various grade levels. In some states, these tests are "high-stakes" events that determine whether a student is promoted to the next grade level in school.

The emphasis on competency tests has grown directly from the national push for higher educational standards and accountability. Under increasing pressure from political leaders, business people, and the general public, policymakers have turned to testing as a primary way to measure and improve student performance. Although experienced educators know that such test results can reveal only a small part of a much broader educational picture, state-mandated competency tests have gained a strong foothold. Teachers must find effective ways to help their students prepare for these tests—and this is where *TestSMART*® plays an important role.

What's inside this book?

Designed to help students review and practice important math and test-taking skills, *TestSMART*® includes reproducible practice exercises in the following areas—

- pretests for each of the five major objectives addressed in the book
- practice exercises that target the specific skills tested within each objective

In addition, each *TestSMART*® book includes—

- a master skills list based on mathematics standards of several states
- complete answer keys

©ECS Learning Systems, Inc.

All rights reserved

The content of *TestSMART®* is outlined below.

Major Objectives: This book focuses on five major mathematics objectives that represent broad areas of understanding generally common to all grade levels. These objectives focus on developing students' understanding of—

- using operations to solve problems
- estimating solutions
- solution strategies/analyzing or solving problems
- mathematical representation
- reasonableness

Specific Skills: A list of specific skills appears below each major objective. These skills represent appropriate grade-level expectations for a given objective.

Pretests: Five pretests are included in this book. Each pretest addresses one of the major objectives and includes test items for all the specific skills included with that objective. Teachers use the pretests to diagnose the students' areas of strength and weakness for a given objective.

Practice Exercises: Practice exercises follow each pretest. Unlike the pretests, each practice exercise addresses a specific skill. Teachers use the practice exercises to target specific areas of weakness revealed in the pretests.

Master Skills List/Correlation Chart: The mathematics skills addressed in *TestSMART®* are based on the mathematics standards and/or test specifications from several different states. No two states have identical wordings for their skills lists, but there are strong similarities from one state's list to another. The Master Skills List for mathematics (page 8) represents a synthesis of the mathematics skills emphasized in various states. Teachers who use this book will recognize the skills that are stressed, even though the wording of a few objectives may vary slightly from that found in their own state's test specifications.

The Master Skills Correlation Chart (page 9) offers a place to identify the skills common to both *TestSMART®* and a specific state competency test. To show how such a correlation can be done, the author has included a sample correlation, which shows both the skills addressed in *TestSMART®* and the skills tested on the Texas Assessment of Knowledge and Skills (TAKS).

Answer Keys: Complete answer keys appear on pages 117–121.

Teaching Tips: The suggested teaching tips and activities found on pages 123–130 offer options usually not found in textbooks.

©ECS Learning Systems, Inc. All rights reserved

How to Use This Book

Effective Test Preparation: What is the most effective way to prepare students for any mathematics competency test? Experienced educators know that the best test preparation includes three critical components—

- a strong curriculum that includes the content and skills to be tested
- effective and varied instructional methods that allow students to learn content and skills in many different ways
- targeted practice that familiarizes students with the specific content and format of the test they will take

Obviously, a strong curriculum and effective, varied instructional methods provide the foundation for all appropriate test preparation. Contrary to what some might believe, merely "teaching the test" performs a great disservice to students. Students must acquire knowledge, practice skills, and have specific educational experiences that can never be included on tests limited by time and in scope. For this reason, books like *TestSMART®* should **never** become the heart of the curriculum or a replacement for strong instructional methods.

Targeted Practice: *TestSMART®* does, however, address the final element of effective test preparation (targeted test practice) in the following ways—

- *TestSMART®* familiarizes students with the content usually included in competency tests.
- *TestSMART®* familiarizes students with the general format of such tests.

When students become familiar with both the content and the format of a test, they know what to expect on the actual test. This, in turn, improves their chances for success.

Using TestSMART®: Used as part of the regular curriculum, *TestSMART®* allows teachers to—

- pretest skills needed for the actual test students will take
- determine students' areas of strength and/or weakness
- provide meaningful test-taking practice for students
- ease students' test anxiety
- communicate test expectations and content to parents

©ECS Learning Systems, Inc.

All rights reserved

Master Skills List

I. The student will use the operations of addition, subtraction, multiplication, and division to solve problems.

A. Use addition, subtraction, multiplication, and division to solve problems involving integers, fractions, and decimals, including multi-step problems

B. Use multiplication and division to find unit rates and ratios in proportional relationships such as speed, density, price, and recipes

II. The student will estimate solutions to a problem situation.

A. Estimate solutions to application problems involving ratios and proportions, including scaling and unit costs

B. Estimate solutions to application problems involving percent, including discounts, markups, commissions, and interest

C. Estimate solutions to application problems involving measurement

III. The student will determine solution strategies and will analyze or solve problems.

A. Solve application problems involving ratios and proportions, including scaling and unit costs

B. Solve application problems involving percent, including discounts, markups, commissions, and interest

C. Identify missing or extraneous information in application problems

IV. The student will express or solve problems using mathematical representation.

A. Represent multiplication and division situations involving fractions and decimals

B. Formulate a possible problem situation when given a simple equation

C. Solve problems by reading and interpreting collected data

V. The student will evaluate the reasonableness of a solution to a problem.

A. Determine the reasonableness of a solution to a problem

©ECS Learning Systems, Inc. All rights reserved

Master Skills Correlation Chart

Use this chart to identify the *TestSMART*® skills included on a specific state competency test. To correlate the *TestSMART*® skills to a specific state's objectives, find and mark those skills common to both. The first column shows a sample correlation based on the Texas Assessment of Knowledge and Skills (TAKS).

	Sample Correlation	
I. The student will use the operations of addition, subtraction, multiplication, and division to solve problems.		
A. Use addition, subtraction, multiplication, and division to solve problems involving integers, fractions, and decimals, including multi-step problems	*	
B. Use multiplication and division to find unit rates and ratios in proportional relationships such as speed, density, price, and recipes	*	
II. The student will estimate solutions to a problem situation.		
A. Estimate solutions to application problems involving ratios and proportions, including scaling and unit costs	*	
B. Estimate solutions to application problems involving percent, including discounts, markups, commissions, and interest	*	
C. Estimate solutions to application problems involving measurement	*	
III. The student will determine solution strategies and will analyze or solve problems.		
A. Solve application problems involving ratios and proportions, including scaling and unit costs	*	
B. Solve application problems involving percent, including discounts, markups, commissions, and interest	*	
C. Identify missing or extraneous information in application problems	*	
IV. The student will express or solve problems using mathematical representation.		
A. Represent multiplication and division situations involving fractions and decimals	*	
B. Formulate a possible problem situation when given a simple equation	*	
C. Solve problems by reading and interpreting collected data	*	
V. The student will evaluate the reasonableness of a solution to a problem.		
A. Determine the reasonableness of a solution to a problem	*	

©ECS Learning Systems, Inc. All rights reserved

Notes

©ECS Learning Systems, Inc.

All rights reserved

Using Operations to Solve Problems

I. Use the operations of addition, subtraction, multiplication, and division to solve problems.

A. Use addition, subtraction, multiplication, and division to solve problems involving integers, fractions, and decimals, including multi-step problems

B. Use multiplication and division to find unit rates and ratios in proportional relationships such as speed, density, price, and recipes

©ECS Learning Systems, Inc. All rights reserved

Objective 1: Pretest

I.A Use addition, subtraction, multiplication, and division to solve problems involving integers, fractions, and decimals, including multi-step problems (1–12)

1. The manager of a store buys school notebooks for 85¢ each and sells them for $1.25. How much would the manager earn if he purchased and sold 15 of the notebooks?

A $12.75

B $7.50

C $6.00

D $5.00

2. The chart shows how many students Mrs. Berry taught at Ruthers Middle School during 4 different school years.

School Year	Number of Students
1998–1999	104
1999–2000	112
2000–2001	98
2001–2002	135

What is the total number of students Mrs. Berry taught during the 4 different school years?

A 549

B 449

C 339

D 239

3. A recipe calls for the following spices:

$\frac{2}{3}$ teaspoon of garlic powder

$\frac{1}{4}$ teaspoon of onion powder

$\frac{3}{4}$ teaspoon of chili powder

$\frac{2}{3}$ teaspoon of salt

What is the total amount of spices needed for the recipe?

A $1\frac{2}{3}$ tsp

B $1\frac{3}{4}$ tsp

C 2 tsp

D $2\frac{1}{3}$ tsp

4. Last year, Mack could run 1 mile in 5.25 minutes. This year, he improved his time by 0.8 minutes. This year, Mack can run 1 mile in—

A 6.05 min

B 5.17 min

C 4.65 min

D 4.45 min

©ECS Learning Systems, Inc.

All rights reserved

5. Each student in Mr. Messmer's class used 16 sheets of typing paper and 4 pieces of construction paper for a booklet. There are 32 students in the class. What is the total number of sheets of both papers the students used?

A 640

B 540

C 512

D 384

6. Edie bought 2 bags of candy. One bag held 350 pieces of candy, and the other held 258 pieces of candy. Edie divided all the candy equally among 8 of her friends. How many pieces of candy did she give each friend?

A 74

B 76

C 78

D 80

7. Todd spent a total of 2 hours and 15 minutes on his science, math, and social studies homework. He spent an equal amount of time on each subject. How much time did Todd spend on his math homework?

A 35 min

B 40 min

C 43 min

D 45 min

8. Melanie's kitten eats $\frac{3}{4}$ cup of cat food every day. How many cups of cat food does the kitten eat in 1 week (7 days)?

A $2\frac{1}{2}$ c

B $4\frac{3}{4}$ c

C $5\frac{1}{4}$ c

D $5\frac{1}{2}$ c

9. Mrs. McKenna had a piece of material that was $4\frac{1}{2}$ yards long. If she cut the material into 5 equal lengths, how long was each piece?

A $\frac{4}{5}$ yd

B $\frac{9}{10}$ yd

C 1 yd

D $1\frac{1}{3}$ yd

10. Meredith works at a restaurant and earns $7.50 per hour. Last week she worked for 22.5 hours. How much money did Meredith earn last week?

A $168.75

B $168.55

C $167.75

D $167.55

©ECS Learning Systems, Inc.

All rights reserved

11. Jerry walked a total of 24.5 miles in 5 days. What was the average distance Jerry walked in 1 day?

A 4.5 mi

B 4.7 mi

C 4.9 mi

D 5 mi

12. A farmer bought 2 bags of grain. One bag weighed 26.3 pounds, and the other bag weighed 26.5 pounds. First he combined all the grain. Then he separated the grain, putting the same amount in each of 8 small bins. If he used all the grain in the 2 bags, how much did he put into each bin?

A 7.6 lb

B 7.2 lb

C 6.8 lb

D 6.6 lb

I.B ***Use multiplication and division to find unit rates and ratios in proportional relationships such as speed, density, price, and recipes (13–24)***

13. A grocery store sells 5 pounds of potatoes for $1.85. How much would you pay for 3 pounds of potatoes?

A $ .91

B $1.11

C $3.05

D $3.08

14. A cookie recipe calls for 24 ounces of flour and will make 40 cookies. Assuming each cookie is about the same size, about how many ounces of flour will be in each cookie?

A 1.6 oz

B 0.7 oz

C 0.6 oz

D 2.9 oz

15. The Scorpions played in 20 soccer games last year and scored a total of 58 goals. What was the average number of goals scored in each game?

A 3.6

B 3.1

C 3

D 2.9

16. Last week, Mrs. Tamio drove 8 hours to her sister's house in another town. Her average speed throughout the whole trip was 67.5 miles per hour. How far did Mrs. Tamio drive in 8 hours?

A 540 mi

B 520 mi

C 516 mi

D 512 mi

17. Gordon paid $248.50 for 5 tires. If he paid the same amount for each tire, how much did each tire cost?

A $59.70

B $49.70

C $47.70

D $41.70

©ECS Learning Systems, Inc.

All rights reserved

18. A juice recipe calls for $3\frac{1}{2}$ cups of sugar and will serve 60 people. How much sugar would you need if you reduced the recipe to serve only 48 people?

A $4\frac{2}{3}$ c

B $3\frac{1}{2}$ c

C $2\frac{4}{5}$ c

D $2\frac{1}{2}$ c

19. A package of watermelon seeds costs $3.50, including tax. There are 25 seeds in each packet. What is the cost of each seed?

A 7¢

B 10¢

C 12¢

D 14¢

20. Mr. Reed bought tickets for 7 teachers to attend a concert. He paid a total of $245 for the tickets. If each ticket cost the same amount, how much did Mr. Reed pay for each ticket?

A $24.50

B $25.00

C $32.00

D $35.00

21. Carmen spent $1.28 for 8 pieces of candy. How much would Carmen have to pay for 15 pieces of candy?

A $3.40

B $2.40

C $2.10

D $1.65

22. Juan's little brother has a small wading pool that holds 64 liters of water. After Juan emptied and cleaned the pool, he refilled it in 12.8 minutes. If he filled the pool at a steady rate, how many liters of water did Juan put in the wading pool each minute?

A 5 L

B 0.5 L

C 0.2 L

D 0.02 L

23. A package holds 60 cookies and costs $4.80. How much does each cookie in the package cost?

A 8¢

B 9¢

C 18¢

D 80¢

24. At a clothing store, Reggie paid $64.48 for 8 shirts. How much would he have paid for 12 of the same shirts?

A $106.72

B $103.20

C $96.72

D $95.62

©ECS Learning Systems, Inc.

All rights reserved

Practice 1.A1

I.A Use addition, subtraction, multiplication, and division to solve problems involving integers, fractions, and decimals, including multi-step problems

1. Francesca spent the following amounts of money at different stores while shopping: \$45.67, \$2.34, \$35.60, \$12.29, \$14.56. How much money did Francesca spend in all?

   A \$112.46

   B \$110.46

   C \$109.46

   D \$95.90

2. Rob and Christy ate an entire pizza that had 10 slices. If Rob ate $\frac{3}{5}$ of the pizza, how many slices of pizza did Christy eat?

   A 4

   B 5

   C 6

   D 7

3. Marian's plant needs $\frac{3}{4}$ cup of water every 3 days. How much water will Marian's plant need over a period of 21 days?

   A $15\frac{3}{4}$ c

   B $5\frac{1}{2}$ c

   C $5\frac{1}{4}$ c

   D 5 c

4. Dominic earned \$14.25 an hour. He worked a total of 22 hours in 3 days, and then he worked an additional 18 hours over 2 more days. How much money did Dominic earn over the 5-day period?

   A \$313

   B \$569

   C \$570

   D \$575

5. A vegetable gardener divides his garden into 5 sections. He uses $\frac{2}{5}$ of the garden for beans, $\frac{1}{10}$ for watermelons, $\frac{1}{6}$ for tomatoes, and the remaining area for cucumbers. What fraction of the garden does the vegetable gardener devote to cucumbers?

   A $\frac{1}{4}$

   B $\frac{1}{3}$

   C $\frac{2}{5}$

   D $\frac{1}{5}$

©ECS Learning Systems, Inc.

All rights reserved

Practice 1.A2

I.A Use addition, subtraction, multiplication, and division to solve problems involving integers, fractions, and decimals, including multi-step problems

1. Sandra normally needs $4\frac{1}{2}$ cups of water to make rice. She decides to triple the recipe when making a rice dish for a dinner party. How much water will she need then?

A $13\frac{1}{2}$ c

B 13 c

C $12\frac{1}{4}$ c

D 12 c

2. A book author had 5 different editors look at parts of her new book. She organized all their comments on the chart below.

Editors' Comments

Type of Comments	Number of Comments
Positive Comments	36
Confusing Section	12
Check Word Choice	22
Punctuation Error	28
Spelling Error	20
Grammar Error	17

If each editor made an equal number of comments, how many comments did each editor make?

A 20

B 25

C 27

D 29

3. To get from work to home, Mrs. Carter spent 2.3 minutes waiting at the bus stop, 34.6 minutes on the bus, and 3.45 minutes walking the rest of the way home. How much time did it take Mrs. Carter to get from work to home?

A 41.35 min

B 40.35 min

C 39.35 min

D 37.35 min

4. Ravi and his 4 friends want to buy 6 bags of lemon drops that cost 55¢ per bag. If Ravi and his friends split the cost equally, how much must each person pay for the bags of lemon drops before sales tax?

A 60¢

B 66¢

C 68¢

D 82¢

5. Mr. Ryan spent 45% of his savings on a down payment for a car and 10% of his savings to repair his computer. If Mr. Ryan had $8,525 in his savings in the beginning, how much money did he have left after making the down payment and repairing the computer?

A $5,967.50

B $4,688.75

C $3,936.75

D $3,836.25

©ECS Learning Systems, Inc.

All rights reserved

Practice 1.A3

I.A Use addition, subtraction, multiplication, and division to solve problems involving integers, fractions, and decimals, including multi-step problems

1. A store owner spent 15¢ each for 35 hair clips, but only sold 17 of them. If the store owner sold each hair clip for $1.99, what was his total profit on the hair clips?

A $5.25

B $27.58

C $28.58

D $33.83

2. Mike runs 2 miles in 11 minutes and 25 seconds. If he continues to run at the same rate, how long will it take him to run 8 miles?

A 44 min 10 sec

B 44 min 40 sec

C 45 min 40 sec

D 46 min

3. Felicia usually buys 12 rolls of paper towels for $5.76. Last week, she bought a new brand that offered 6 rolls for $2.40. How much money did Felicia save on each roll of paper towels?

A 40¢

B 12¢

C 9¢

D 8¢

4. Jennifer's recipe for hot chocolate calls for $\frac{1}{4}$ cup milk and $\frac{1}{3}$ cup water per serving. How much water is required if she wants to make 14 servings?

A $5\frac{1}{3}$ c

B $4\frac{2}{3}$ c

C $4\frac{1}{3}$ c

D $3\frac{1}{2}$ c

5. The annual birth rate in Asia is approximately 27 births for every 1,000 people. If there are currently 3,400,000,000 people in Asia, how many people will be born there this year?

A 918,000

B 91,800,000

C 92,200,000

D 93,800,000

6. Jeremiah wants to buy an iguana that costs $24.95. He has saved $12.35 and will earn $5.00 in allowance this week. After receiving his allowance, how much more money will Jeremiah need to buy the iguana?

A $12.60

B $8.20

C $7.80

D $7.60

©ECS Learning Systems, Inc.

All rights reserved

Practice 1.A4

I.A Use addition, subtraction, multiplication, and division to solve problems involving integers, fractions, and decimals, including multi-step problems

1. The chart shows how many bottle caps Sean collected in 4 years.

Year	Number of Bottle Caps
1998	56
1999	124
2000	95
2001	165

What is the total number of bottle caps that Sean collected during the 4-year period?

A 384

B 400

C 420

D 440

2. Mrs. Ortiz uses $\frac{1}{3}$ bottle of cooking oil each time she makes a large batch of pasta for a restaurant. She has 24 bottles of cooking oil in stock. If she makes a large batch of pasta each day for 12 days in a row, how many bottles of cooking oil will she have left in stock?

A 20

B 18

C 15

D 8

3. Molly planted $\frac{1}{2}$ package of bluebonnet seeds and $\frac{1}{3}$ package of daisy seeds in her yard. She saved the remaining seeds. How many packets of flower seeds will she have left to plant next year?

A $\frac{1}{6}$

B $\frac{5}{6}$

C 1

D $1\frac{1}{6}$

4. Mr. Parks purchased a car that cost $12,986.40. He made equal payments each month for 4 years to pay it off. How much were Mr. Parks' monthly payments? (12 months = 1 year)

A $1,082.20

B $280.60

C $270.55

D $260.65

5. Theo and Clint are roommates. Clint pays 56% of the $825 rent each month, and Theo pays for the remainder of the rent plus the water bill. How much money does Theo pay for rent each month?

A $360.80

B $363.00

C $407.00

D $462.00

©ECS Learning Systems, Inc.

All rights reserved

Practice 1.A5

I.A Use addition, subtraction, multiplication, and division to solve problems involving integers, fractions, and decimals, including multi-step problems

1. Daniel saved $4.30 each week for a new telephone that cost $34.90. Daniel also had to pay a 10% sales tax on the phone. How many weeks did Daniel have to save in order to have enough money to buy the phone?

   A 7

   B 8

   C 9

   D 10

2. Caitlyn bought 3 kinds of soup for a casserole recipe. She used $\frac{3}{4}$ can of cream of celery soup, $\frac{4}{5}$ can of cream of mushroom soup, and $\frac{1}{2}$ can of chicken broth. How many cans of soup will she use if she makes the recipe twice?

   A $2\frac{1}{20}$ cans

   B $3\frac{1}{2}$ cans

   C $4\frac{1}{20}$ cans

   D $4\frac{1}{10}$ cans

3. Mina scored 25% of her team's 32 points during a basketball game. She scored 40% of her team's 45 points during the next game. What was the total number of points Mina scored in the 2 games?

   A 8

   B 10

   C 18

   D 26

4. Mr. Cree sold 35 soccer tickets for $20 apiece. He earned 22% of his total sales. How much money did he earn from selling the tickets?

   A $280.00

   B $154.00

   C $148.00

   D $12.10

5. At Miller Middle School, $\frac{2}{3}$ of the students took part in field day. Of those students, $\frac{1}{2}$ won medals. What fraction of the school's students won medals at field day?

   A $\frac{1}{6}$

   B $\frac{1}{3}$

   C $\frac{1}{2}$

   D $\frac{2}{3}$

©ECS Learning Systems, Inc. All rights reserved

Practice 1.A6

I.A Use addition, subtraction, multiplication, and division to solve problems involving integers, fractions, and decimals, including multi-step problems

1. Fredrick built a small cabinet by cutting a large piece of wood into 4 smaller pieces that were $2\frac{1}{4}$ meters, $\frac{3}{4}$ meter, $\frac{5}{6}$ meter, and $1\frac{1}{3}$ meters long. How long was the large piece of wood originally?

   A $5\frac{1}{2}$ m

   B $5\frac{1}{3}$ m

   C $5\frac{1}{6}$ m

   D $4\frac{5}{6}$ m

2. Kelly has a rectangular pane of glass that measures 9 inches by 12 inches. He wants to place the pane in a rectangular space that measures 100 square inches. How much smaller must the pane of glass be to fit in the space?

   A 8 in^2

   B 10 in^2

   C 12 in^2

   D 58 in^2

3. Prithi purchased 4 blouses that cost $12.85 each, including tax. She gave the cashier $60. How much change did she receive?

   A $43.15

   B $8.80

   C $8.60

   D $8.40

4. A new television program had 3,200,000 viewers for its first episode. There were 350,000 fewer viewers for the program's second episode. The number of viewers dropped another 175,000 for the third episode. How many viewers watched the third episode?

   A 2,850,000

   B 2,675,000

   C 2,575,000

   D 525,000

5. Brandon earns 27¢ for every $3 worth of car parts that he sells. If Brandon sells $51 worth of car parts, how much money will he earn?

   A $9.00

   B $5.59

   C $4.79

   D $4.59

6. Mr. Nguyen bought a bouquet of flowers that was $\frac{1}{2}$ roses, $\frac{1}{3}$ daisies, and $\frac{1}{6}$ lilies. There were 18 flowers in the bouquet. How many were lilies?

   A 9

   B 6

   C 4

   D 3

©ECS Learning Systems, Inc.

All rights reserved

Practice 1.A7

I.A Use addition, subtraction, multiplication, and division to solve problems involving integers, fractions, and decimals, including multi-step problems

1. A racecar driver circled the racetrack 5 times during warm-up. His times for each lap were: 45.6 seconds, 39.9 seconds, 41.5 seconds, 42.7 seconds, and 40.5 seconds. How much time did he spend on the 5 laps during warm-up?

   A 228 sec

   B 220.2 sec

   C 211.2 sec

   D 210.2 sec

2. Mrs. Ashby's class volunteered to plant trees in the nearby park. The park was 8.5 acres, and the volunteers decided to plant 12 trees per acre. How many trees did they need for the entire park?

   A 103

   B 102

   C 101

   D 20

3. On Monday, Julia studied for 1.2 hours. The next day she studied twice as long. She also watched television for 2 hours. How much more time did Julia spend studying over the 2 days than watching television?

   A 1.6 hr

   B 1.8 hr

   C 2.6 hr

   D 9.6 hr

4. Mrs. Madison sent a job application to 35 different businesses and received a response from $\frac{2}{5}$ of them. How many of the businesses did **NOT** respond?

   A 10

   B 14

   C 21

   D 25

5. During the first 2 days of school, students at Monroe Middle School could sign up to be office aids. On the first day, 108 students signed up. On the second day, 138 students signed up. If the school secretary wants the students to work in teams of 6, how many teams could be formed from the students who signed up on both days?

   A 41

   B 36

   C 23

   D 18

6. Salvador earned $28 per day for 7 days while mowing lawns. He then earned $8 for walking his neighbor's dogs. How much money did he earn in all?

   A $194

   B $196

   C $204

   D $206

©ECS Learning Systems, Inc. All rights reserved

Practice 1.B1

I.B Use multiplication and division to find unit rates and ratios in proportional relationships such as speed, density, price, and recipes

1. Mrs. Schubert can enter text on a computer at a rate of 85 words every 2 minutes. If she has to type an essay that is 1,530 words long, how long will it take her to type it?

   A 38 min

   B 36 min

   C 24 min

   D 18 min

2. Desmond took 3 hours 18 minutes to complete his test. He spent the same amount of time on each section of the test. If there were 6 sections on the test, how much time did Desmond spend on each section?

   A 28 min

   B 30 min

   C 33 min

   D 35 min

3. On a business trip, Ernesto traveled an average of 65.5 miles per hour. He was in the car for 5.5 hours. How far did he travel?

   A 11.91 mi

   B 36.025 mi

   C 360.05 mi

   D 360.25 mi

4. Allison took 4 rolls of film to be developed at a store. All 4 rolls had 24 exposures. If Allison paid $24 to have all 4 rolls developed, how much did she pay per picture?

   A $0.25

   B $1

   C $4

   D $24

5. At the produce market, Sally can buy 32 yams for $5.76. If Sally wants to buy 52 yams, how much will she pay?

   A $6.28

   B $8.36

   C $9.26

   D $9.36

6. A 4-person relay team ran 1 mile in 4 minutes 32 seconds. What is the average time it took each runner to run his lap?

   A 58 sec

   B 1 min

   C 1 min 8 sec

   D 1 min 10 sec

©ECS Learning Systems, Inc.

All rights reserved

Practice 1.B2

***I.B** Use multiplication and division to find unit rates and ratios in proportional relationships such as speed, density, price, and recipes*

1. Kayla can make 6 kitchen aprons in 12 hours 30 minutes. How long does it take her to make 1 apron?

   A 1 hr 50 min

   B 2 hr

   C 2 hr 5 min

   D 2 hr 15 min

2. When Mitch makes lasagna for 12 people, he uses 16 ounces of pasta. How many ounces of pasta will he use if he makes lasagna for 3 people?

   A 2 oz

   B 4 oz

   C 6 oz

   D 8 oz

3. DeAnn's car holds 16 gallons of gasoline. Her car uses gasoline at a rate of 1.4 gallons per 28 miles. If she begins with a full tank of gasoline, how many miles can she travel before her car's gas tank is empty?

   A 290 mi

   B 320 mi

   C 340 mi

   D 627.2 mi

4. Dr. Bratcher ordered 14 cases of medical supplies that weighed 6.7 pounds each and cost a total of $969.50. How much did one case of supplies cost?

   A $144.70

   B $72.25

   C $70.25

   D $69.25

5. A machine at a chocolate factory can fill 1 box of chocolates in 2.6 minutes. How many boxes of chocolate can it fill in 1,001 minutes?

   A 385

   B 383

   C 380

   D 375

6. A plane traveled 2,340 miles in 6 hours to reach its destination on time. If the plane traveled at a constant speed, how far did it travel in 2 hours?

   A 790 mi

   B 780 mi

   C 760 mi

   D 390 mi

©ECS Learning Systems, Inc. All rights reserved

Practice 1.B3

I.B Use multiplication and division to find unit rates and ratios in proportional relationships such as speed, density, price, and recipes

1. The width of Sinead's wall is 19.2 feet. She wants to cover the wall with wallpaper, which comes on rolls that are 3.2 feet wide. How many strips of wallpaper must Sinead use to completely cover the wall?

   A 4

   B 5

   C 6

   D 7

2. Valerie has 14 charms on her bracelet. Together, the charms and bracelet are worth $124.50. If the bracelet costs $12.50, what is the average cost of each charm?

   A $8.00

   B $8.80

   C $8.90

   D $9.96

3. Mr. Peters, a florist, knows that he usually sells 3 potted plants for every 11 bouquets of flowers he sells. If he sold 45 potted plants in 2 weeks, how many bouquets of flowers did he probably sell during the same time period?

   A 167

   B 165

   C 135

   D 30

4. A recipe for 1 loaf of bread calls for 0.3 pound of butter. A bakery uses 1.2 pounds of butter in every large batch of bread dough. How many loaves does each large batch of bread dough make?

   A 40

   B 6

   C 4

   D 3

5. A candlemaker adds no more than 0.2 ounces of scented oil to any small wax candle she makes. If the candlemaker has 14 ounces of scented oil, how many small wax candles can she make?

   A 7

   B 14

   C 60

   D 70

6. An architect used a total of 48 pages to print complete blueprints for several new homes. Each complete blueprint covered 3 pages. How many complete blueprints did the architect print?

   A 16

   B 18

   C 144

   D 160

©ECS Learning Systems, Inc. All rights reserved

Practice 1.B4

I.B Use multiplication and division to find unit rates and ratios in proportional relationships such as speed, density, price, and recipes

1. There are several small flower beds scattered across the grounds of a museum. There are 2 flower beds for every 40 square feet of land. If there is a total of 65 flower beds, how large are the museum grounds?

   A 2,925 ft^2

   B 1,350 ft^2

   C 1,300 ft^2

   D 1,200 ft^2

2. A boat traveled 122.4 miles in 3.4 hours. If the boat traveled at a steady rate, how far did it travel in 2 hours?

   A 144 mi

   B 72 mi

   C 61.2 mi

   D 36 mi

3. Mrs. Mazur is a court reporter. If she can type 152 words per minute, how many words can she type in 1 hour and 30 minutes?

   A 228

   B 12,160

   C 13,580

   D 13,680

4. A farmer mixes 1.5 cups of dried corn together with every 6 cups of regular chicken feed. How much dried corn will a farmer add to a bag that contains 78 cups of regular chicken feed?

   A 52 c

   B 19.5 c

   C 18.5 c

   D 13 c

5. A restaurant charged $1,927.41 for catering a formal dinner. If 123 people attended the dinner, what was the average cost of each person's meal?

   A $14.67

   B $15.57

   C $15.67

   D $16.67

6. During 4 weeks of work, a hair stylist uses 3 bottles of shampoo. If the hair stylist worked 46 weeks last year, how many bottles of shampoo did she use?

   A 34.5

   B 35.5

   C 37

   D 61.3

©ECS Learning Systems, Inc. All rights reserved

Practice 1.B5

I.B Use multiplication and division to find unit rates and ratios in proportional relationships such as speed, density, price, and recipes

1. Mason worked 53 hours in 1 work week (5 days). What is the average amount of time he worked each day?

   A 5.3 hr

   B 9.1 hr

   C 10.6 hr

   D 10.8 hr

2. Sharon spends \$10.50 to wash and dry 7 loads of laundry. What is the average cost of washing and drying 1 load of laundry?

   A 75¢

   B \$1.50

   C \$1.75

   D \$2.00

3. Joli heated water in the science lab. When she started, the water's temperature was 15˚C. After 12 minutes of heating, the water's temperature was 78˚C. Assuming the water's temperature rose at a steady rate, at what rate did the temperature increase per minute?

   A 6.5˚C

   B 5.3˚C

   C 5.25˚C

   D 5.15˚C

4. Tyler drove for 1 hour 30 minutes. He traveled a distance of 105 miles. What was Tyler's average speed?

   A 60 mph

   B 65 mph

   C 68 mph

   D 70 mph

5. Kim used a whole box of tissues in 3 weeks. If the box contained 252 tissues, what was the average number of tissues Kim used each day?

   A 12

   B 14

   C 62

   D 66

6. Mrs. Sergio earns \$1,285 every 2 weeks. The company deducts \$257 from each pay check for taxes and insurance. How much does Mrs. Sergio pay for taxes and insurance for 1 full year (52 weeks)?

   A \$13,364

   B \$6,682

   C \$6,027

   D \$3,084

©ECS Learning Systems, Inc.

All rights reserved

Practice 1.B6

I.B ***Use multiplication and division to find unit rates and ratios in proportional relationships such as speed, density, price, and recipes***

1. Turkey has an estimated population of 62,032,000. The population density of Turkey is 80 people per square kilometer. What is the estimated area (in square kilometers) of Turkey?

   A 46,625,600 km^2

   B 785,400 km^2

   C 775,400 km^2

   D 77,540 km^2

2. Airtime Airlines charges $264.60 for a one-way ticket from San Diego, California, to Phoenix, Arizona. If it is 352.8 miles from San Diego to Phoenix, what is the cost per mile to travel from San Diego to Phoenix?

   A 70¢

   B 75¢

   C 85¢

   D $1.33

3. Grade-A Grocery Store sells 3 apple pies for $4.59. How much would it cost to buy 7 pies?

   A $6.55

   B $7.91

   C $10.51

   D $10.71

4. Keith likes to have extra shoelaces in case one breaks. He buys 3 pairs of shoelaces for every 2 pairs of shoes he owns. How many pairs of shoelaces should Keith buy if he owns 14 pairs of shoes?

   A 7

   B 19

   C 21

   D 24

5. Susan goes hiking 10 times per year. She can hike a mountain at a rate of 1.8 miles per hour. If Susan decides to spend 4.75 hours hiking up a mountain, how far will she go?

   A 10.8 mi

   B 8.55 mi

   C 8.45 mi

   D 6.55 mi

6. Last month, Mr. and Mrs. Fratelli spent $60.92 at restaurants. If they ate at restaurants 4 times last month, what was the average cost of each restaurant bill?

   A $12.73

   B $15.13

   C $15.23

   D $30.46

©ECS Learning Systems, Inc. All rights reserved

Practice 1.B7

I.B ***Use multiplication and division to find unit rates and ratios in proportional relationships such as speed, density, price, and recipes***

1. At the end of the school year, Mr. Higgins looked at the grades his students had earned. He discovered that the students earned 4 A's for every 9 B's they earned. If the students earned a total of 1,296 B's, how many A's did they earn?

A 144

B 324

C 576

D 674

2. A fire regulation states that an arena should seat no more than 12 people for every 100 square feet of floor space. If an arena has a total of 36,000 square feet of floor space, what is the greatest number of people that should be seated in the arena?

A 3,000

B 4,220

C 4,310

D 4,320

3. A female spider can lay 3,450 eggs when it is ready to reproduce. Of the 3,450 eggs, 69 survive to become baby spiders. If a spider lays 27,600 eggs in its lifetime, how many baby spiders will it produce?

A 400

B 500

C 542

D 552

4. A copy machine's temperature increases at a rate of 0.01°F for each copy it makes in a row. If the machine makes 451 copies in a row, by how much will its temperature increase?

A 4.51°F

B 4.61°F

C 5.51°F

D 45.1°F

5. Mrs. Still's babysitter charges by the hour. Last week, Mrs. Still paid the babysitter $49.50 for 9 hours of work. What was the babysitter's hourly rate?

A $5.75

B $5.60

C $5.55

D $5.50

6. Brayton played a video game with his 3 friends. He earned the highest score 2 out of every 9 times that he and his friends played. If all 4 friends played the game 36 times, how many times did Brayton get the highest score?

A 18

B 12

C 8

D 4

©ECS Learning Systems, Inc. All rights reserved

Notes

©ECS Learning Systems, Inc. All rights reserved

Estimating Solutions

II. Estimate solutions to a problem situation.

A. Estimate solutions to application problems involving ratios and proportions, including scaling and unit costs

B. Estimate solutions to application problems involving percent, including discounts, markups, commissions, and interest

C. Estimate solutions to application problems involving measurement

©ECS Learning Systems, Inc.

All rights reserved

Objective 2: Pretest

II.A Estimate solutions to application problems involving ratios and proportions, including scaling and unit costs (1–10)

1. There were 26 women and 54 men at a local gym. Which is the best estimate for the ratio of men to women at the gym?

   A 2 to 2

   B 2 to 1

   C 1 to 3

   D 1 to 2

2. Joanna works in the school book-store every day for 90 minutes. On Monday, she sold 18 pencils and 5 packages of notebook paper. If she sells about the same number of pencils every day, which is the best estimate for the number of pencils she would sell during 2 full weeks of school (10 days)?

   A 240

   B 220

   C 200

   D 150

3. A school uses 18 vans to take students on field trips. Each van holds between 9 and 12 students. Last week, the school used all 18 vans to take students to a concert. Which is the best estimate for the number of students who rode in the vans?

   A 300

   B 260

   C 240

   D 200

4. An architect drew a scale model of a fireplace. In the model, the fireplace was 3.8 inches tall and 2.15 inches wide. The actual fireplace was 10.75 feet wide. Which is the best estimate for the height of the actual fireplace?

   A 20 ft

   B 17 ft

   C 15 ft

   D 12 ft

5. Ruth bought 14 tomatoes for $4.39. Which is the best estimate for the average cost of each tomato?

   A 25¢

   B 27¢

   C 30¢

   D 33¢

6. The chart shows the number of boys and girls that attended a school play on 5 different days.

Class	Boys	Girls
Monday	28	15
Tuesday	15	8
Wednesday	31	14
Thursday	53	24
Friday	63	32

   The play was presented to a total of 585 students over a 3-week period of time. Based on the ratio of boys to girls, which is the best estimate for the total number of boys that attended the play during the 3 weeks?

   A 190

   B 250

   C 300

   D 400

©ECS Learning Systems, Inc.

All rights reserved

7. Audra uses $1\frac{1}{2}$ teaspoons of sugar in each 12-ounce glass of iced tea she drinks. Last week, Audra drank an entire pitcher of iced tea. If the pitcher held 82 ounces of tea, which is the best estimate for the amount of sugar Audra used?

   A 8

   B 10

   C 12

   D 14

8. Mike helps the school librarian by shelving returned books. He shelves an average of 18 books in 15 minutes. Which is the best estimate for the total number of books Mike could shelve in 5 days if he worked for 30 minutes each day?

   A 250

   B 225

   C 200

   D 160

9. On a map of the United States, 2 cities are 6.2 centimeters apart. The map is drawn to scale so that 1.5 centimeters represents 100 miles. Which is the best estimate for the actual distance between the 2 cities?

   A 620 mi

   B 520 mi

   C 470 mi

   D 400 mi

10. A merry-go-round makes 3 complete revolutions every 33.25 seconds. Which is the best estimate for how long it would take the merry-go-round to make 25 complete revolutions?

    A 3 min

    B 3.5 min

    C 4 min

    D 4.5 min

©ECS Learning Systems, Inc.

All rights reserved

II.B Estimate solutions to application problems involving percent, including discounts, markups, commissions, and interest (11–20)

11. Miriam wants to buy a coat that costs $45.99. The final cost of the coat will include an 8.25% sales tax. Which is the best estimate for the final price of the coat, including tax?

A $40

B $45

C $50

D $54

12. At Griffin Middle School, 128 seventh-grade students were on the honor roll last semester. If there are a total of 613 seventh-grade students at the school, which is the best estimate for the percent of seventh-grade students that were on the honor roll?

A 25%

B 20%

C 15%

D 10%

13. Mr. Singh works at a car dealership. He earns a 3.2% commission on his total sales each year. Last year, Mr. Singh's total sales equalled $1,028,500. Which is the best estimate for the commission Mr. Singh earned last year?

A $45,000

B $30,000

C $27,000

D $10,000

14. The manager of a grocery store buys 1 gallon of milk for $1.62 and then sells each gallon for $2.49. Which is the best estimate for the percentage of markup on 1 gallon of milk?

A 35%

B 40%

C 45%

D 55%

15. Customers who pay cash at the Top Stop gas station receive a 1.85% discount on their purchases. Which is the best estimate for the amount you would save if you paid cash for purchases that came to $41.15?

A $1.85

B 80¢

C 40¢

D 18.5¢

16. Mr. Herman deducts 2.5% from a student's grade if an assignment is turned in late. Yesterday, Eddie turned in an assignment late. If Eddie had turned it in on time, his grade would have been a 91. Which is the best estimate for Eddie's final grade for the late assignment?

A 80

B 84

C 88

D 90

©ECS Learning Systems, Inc.

All rights reserved

17. Last year, students at Cherry Creek Middle School were absent a total of 1,290 times. This year, student absences increased by 2.8%. Which is the best estimate for the total number of student absences this year?

A 1,350

B 1,475

C 1,525

D 1,650

18. Jill delivers newspapers in her neighborhood. She earns a 3.2% commission on her total sales each month. Last month, her total sales were $4,694. Which is the best estimate for the commission Jill earned last month?

A $1,500

B $180

C $140

D $100

19. On Wednesday, workers at the Great Gizmo Factory made 2,950 gizmos. Of those 2,950 gizmos, 148 were defective and had to be thrown out. Which is the best estimate for the percent of gizmos that were **NOT** defective?

A 95%

B 90%

C 65%

D 5%

20. Mr. Hansen earns $488 per week. His boss has promised him a 4.25% pay raise. Which is the best estimate for the amount Mr. Hansen will earn each week after his pay raise?

A $490

B $520

C $550

D $575

©ECS Learning Systems, Inc. All rights reserved

II.C Estimate solutions to application problems involving measurement (21–30)

21. Mrs. Frantzen made 9 dance costumes for the students in her dance class. She used 2.85 meters of material for each costume. Which is the best estimate for the total amount of cloth she used to make all the costumes?

A 12 m

B 18 m

C 27 m

D 30 m

22. Nelda travels a total of 8.95 miles each day to and from school. Which is the best estimate for the total distance she travels to and from school during a 182-day school year?

A 1,400 mi

B 1,600 mi

C 1,800 mi

D 2,000 mi

23. Each student in a seventh-grade science class needs 9.2 milligrams of a substance to complete an experiment. If a total of 195 students are in this science class, which is the best estimate for the total amount of the substance they will use for the experiment?

A 180 mg

B 1.5 g

C 1.8 g

D 2 g

24. Each student desk in a classroom is 58 centimeters wide. The classroom is 9.75 meters wide. Which is the best estimate for the greatest number of desks you could place side by side across the entire width of the classroom?

A 10

B 12

C 15

D 20

25. Jody made punch for a school party. She bought a bottle that held 3.15 liters of sparkling water. The punch recipe called for 310 milliliters of sparkling water. Which is the best estimate for the total number of times Jody could prepare the recipe with 1 bottle of sparkling water?

A 12

B 10

C 8

D 5

26. David made a model boat. He used 2 strips of balsa wood that were each $8\frac{5}{6}$ inches long. He used 4 strips of pine wood that were each $6\frac{3}{4}$ inches long. Which is the best estimate for the total length of all 6 pieces of wood?

A 3 ft

B 4 ft

C $4\frac{1}{2}$ ft

D 5 ft

©ECS Learning Systems, Inc.

All rights reserved

27. A supervisor timed a work crew that was paving a section of highway and found that they could pave about 18 meters of highway every hour. The crew paved a total of 5.86 kilometers of highway. Which is the best estimate for the amount of time the crew spent on the paving job?

A 200 hr

B 240 hr

C 270 hr

D 300 hr

28. Carrie treated a swimming pool with 4 different chemicals. The chart shows how much of each chemical she poured into the pool.

Chemical	Amount
A	0.38 g
B	255 mg
C	185 mg
D	0.81 g

Which is the best estimate for the total amount of chemicals Carrie poured into the pool?

A 400 mg

B 600 mg

C 1.2 g

D 1.6 g

29. For an earache, Matthew took 24.5 milliliters of medicine 4 times every day. He took the medicine for 8 days. Which is the best estimate for the total amount of medicine Matthew took for the earache?

A 800 mL

B 600 mL

C 400 mL

D 100 mL

30. A spool held 11.8 meters of thread. Mrs. Stern used the following amounts of thread in her sewing project:

97.5 cm
12.25 cm
1.12 m
78.1 cm
1.85 m

Which is the best estimate for the amount of thread left on the spool after Mrs. Stern finished her sewing project?

A 10 m

B 8.5 m

C 7 m

D 6.25 m

©ECS Learning Systems, Inc. All rights reserved

Practice 2.A1

II.A Estimate solutions to application problems involving ratios and proportions, including scaling and unit costs

1. There were 29 female guppies and 61 male guppies in an aquarium at the pet store. Which is the best estimate for the ratio of male to female guppies?

A 1 to 2

B 1 to 3

C 2 to 1

D 2 to 3

2. Leeza bought a package of 48 blank CDs. She recorded information on 10 CDs. Which is the best estimate for the ratio of used CDs to unused CDs?

A 1 to 5

B 1 to 4

C 3 to 1

D 5 to 1

3. Lou bought 35 gumballs for $7.10. Which is the best estimate for the average price of each gumball?

A 20¢

B 22¢

C 25¢

D $4.90

4. Joel needs 26 small nails and 74 medium nails to finish his woodworking project. Which is the best estimate for the ratio of small to medium nails he needs?

A 3 to 1

B 2 to 7

C 1 to 4

D 1 to 3

5. A rock tumbler makes 18 complete revolutions every 54.75 seconds. Which is the best estimate for the amount of time it would take the tumbler to make 60 complete revolutions?

A 2 hr 30 min

B 1 hr 50 min

C 3 min

D 6 min

6. Sophia is a fashion designer. She creates scale drawings of her clothing designs on a scale of 0.7 inches = 6.8 inches. If she designs a skirt that is 3.8 inches long in the drawing, which is the best estimate for the skirt's actual length?

A 40 in

B 26 in

C 18 in

D 0.4 in

7. Juan drives back and forth from work 12 times each week. He travels a total of 456 miles weekly. If Juan can travel 26.8 miles on 1 gallon of gas, which is the best estimate for how many gallons of gas he uses each week traveling back and forth from work?

A 18 gal

B 16 gal

C 13.3 gal

D 2.25 gal

©ECS Learning Systems, Inc.

All rights reserved

Practice 2.A2

II.A Estimate solutions to application problems involving ratios and proportions, including scaling and unit costs

1. Carlos bought 5 packages of gum for \$2.90. Each package held 10 pieces of gum. Which is the best estimate for the average cost of each piece of gum?

   A 60¢

   B 30¢

   C 20¢

   D 6¢

2. Sue bought a package that held 41.5 ounces of fish food. She feeds her fish 3.8 ounces of food each day. Which is the best estimate for the number of days it would take Sue to feed all the food to her fish?

   A 8

   B 10

   C 13

   D 14

3. In a school band, 5 students play the flute and 21 students play the clarinet. Which is the best estimate for the ratio of clarinet players to flute players?

   A 1 to 5

   B 1 to 4

   C 4 to 1

   D 5 to 2

4. Cecilia and her mother ride the train a total of 3.85 hours to visit her grandparents once a month. Which is the best estimate for the total amount of time Cecilia and her mother will spend on the train in 1 year (12 months)?

   A 60 hr

   B 48 hr

   C 40 hr

   D 4 hr

5. On a map of his town, Keith noticed that his school was 5.8 centimeters from the public library. The map is drawn to scale so that 1.2 centimeters represents 5 miles. Which is the best estimate for the actual distance between Keith's school and the public library?

   A 1.2 mi

   B 15 mi

   C 20 mi

   D 30 mi

6. A television station airs 1 public-service announcement for every 4 commercials it airs. If the television station shows 62 commercials every 2 hours, which is the best estimate for how many public-service announcements there will be during the same period of time?

   A 8

   B 12

   C 15

   D 30

©ECS Learning Systems, Inc.

All rights reserved

Practice 2.A3

II.A Estimate solutions to application problems involving ratios and proportions, including scaling and unit costs

1. Margarite is building a scale model of a storage shed. She will use the scale model as a doghouse. The shed is 8.15 feet tall and 5.75 feet wide. Margarite wants the doghouse to be 2 feet tall. Which is the best estimate for the width of the doghouse?

A 2.6 ft

B 1.5 ft

C 1.2 ft

D 1 ft

2. Marcos planted 48 bean plants as part of a science experiment. Of the 48 plants, only 34 grew. Which is the best estimate for the ratio of plants that grew to those that did not grow?

A 1 to 3

B 1 to 4

C 3 to 1

D 3 to 5

3. Joan observed the night sky for 1 hour 30 minutes each night and recorded the number of satellites passing over her area. On the first night, she observed 11 satellites. Which is the best estimate for the number of satellites Joan would expect to see during 15 hours of observation?

A 70

B 100

C 150

D 200

4. The chart shows the number of salmon and tuna caught by a fishing boat on 5 different days. The total number of fish caught was 412.

Day of Week	Salmon	Tuna
Monday	49	41
Tuesday	60	35
Wednesday	53	19
Thursday	71	28
Friday	37	19

Which is the best estimate for the ratio of the total number of salmon caught to the total number of tuna that were caught?

A 7 to 10

B 2 to 1

C 1 to 2

D 1 to 1.5

5. A business bought 9 boxes of paper. Each box contained 10 reams of paper. Which is the best estimate for the number of work days the paper will last if the business uses 1.85 reams of paper each work day?

A 55 days

B 50 days

C 45 days

D 5 days

©ECS Learning Systems, Inc.

All rights reserved

Practice 2.A4

II.A Estimate solutions to application problems involving ratios and proportions, including scaling and unit costs

1. Burt bought a carton of soft drinks at the store for $3.69. The carton held 12 cans of soft drinks. Which is the best estimate for the average cost of each soft drink in the carton?

   A 30¢

   B 34¢

   C 36¢

   D 40¢

2. At Woodland Middle School, there are 118 boys in the seventh grade. Only 13 seventh-grade boys are members of the school basketball team; 39 seventh-grade boys are members of the football team. Which is the best estimate for the ratio of seventh-grade boys on the football team to the total number of seventh-grade boys?

   A 3 to 1

   B 2 to 5

   C 1 to 4

   D 1 to 3

3. Kim wants to save her science project on computer disks. One disk will hold 1.4 megabytes of information. If Kim's project contains 5.8 megabytes of information, which is the best estimate for the total number of disks she will need?

   A 2

   B 3

   C 4

   D 6

4. The students at Conrad Middle School are building a replica of a ship. They are using the design from a toy store model to build the ship. The plans for the model show that the finished ship will be 11 inches tall and 14 inches long. The students want their replica to stand 4 feet tall. If the students build their replica to scale, which is the best estimate for its length?

   A 2.5 ft

   B 3 ft

   C 5 ft

   D 30 ft

5. There are 246 campers at Oak Trails Summer Camp. The cafeteria makes 3 eggs for every 2 campers for breakfast each morning. Which is the best estimate for the number of eggs the cafeteria makes each day?

   A 375

   B 167

   C 150

   D 50

6. Every other day, Devin runs around a circular track 5.75 times in 32 minutes. The distance of the track is 0.34 miles. Which is the best estimate for the number of times Devin circles the track in 8 days?

   A 30

   B 24

   C 7

   D 2.3

©ECS Learning Systems, Inc.

All rights reserved

Practice 2.A5

II.A Estimate solutions to application problems involving ratios and proportions, including scaling and unit costs

1. A farmer built a scale model of a windmill to use as a decoration on his property. The actual windmill was 20 feet tall with blades that were 1.9 feet long. The scale model was 4.75 feet tall. Which is the best estimate for the length of each blade on the scale model?

   A 0.4 ft

   B 0.5 ft

   C 1.5 ft

   D 5 ft

2. Mai can walk 8.25 feet in 3.2 seconds. Which is the best estimate for the distance she can walk in 15 seconds?

   A 6 ft

   B 20 ft

   C 40 ft

   D 45 ft

3. At a track meet, the red team had 5 of its 32 members qualify for the final round of competition. Which is the best estimate for the ratio of team members who qualified for the final competition to those who did not qualify?

   A 1 to 4

   B 1 to 5

   C 1 to 6

   D 6 to 1

4. Mrs. Henry kept track of the pens she sold in her store. For every 3 red pens she sold, she sold 16 blue pens. On Saturday, she sold 11 red pens. Which is the best estimate for the number of blue pens she sold on Saturday?

   A 2

   B 45

   C 50

   D 70

5. Jackson sold candy to raise money for his school. When the sale was over, he had 14 mint candies and 37 chocolate candies left. Which is the best estimate for the ratio of mint to chocolate candies he had left?

   A 2 to 7

   B 2 to 5

   C 1 to 4

   D 1 to 3

6. Vanessa bought 678 balloons to decorate for a school dance. She paid $16.88 for all of the balloons. Which is the best estimate for the cost of each balloon?

   A 2.5¢

   B 3.5¢

   C 4¢

   D 35¢

©ECS Learning Systems, Inc.

All rights reserved

Practice 2.B1

II.B Estimate solutions to application problems involving percent, including discounts, markups, commissions, and interest

1. Samuel wants to buy a video game that costs $24.99. The final cost of the video game will include 7.75% sales tax. Which is the best estimate for the final price of the video game, including tax?

   A $27

   B $25

   C $24

   D $23

2. Hal earns 3.3% interest yearly on the money in his savings account. He currently has $2,519 in his savings account. Which is the best estimate for the amount of money he will earn in interest this year?

   A $65

   B $75

   C $750

   D $833

3. Ginger has a coupon for 15% off her total purchase at a department store. Which is the best estimate for the amount she will save if she buys a sweater that costs $34.97?

   A $52.50

   B $29.75

   C $5.25

   D $4.75

4. Chandra paid a 25% deposit to rent an innertube at the water park. Which is the best estimate for the cost of the innertube if the deposit was $2.63?

   A $10

   B $12

   C $15

   D $62.50

5. A furniture store sells oak dressers for $395. They sell the same dresser at an 8% discount if it is unpainted. Which is the best estimate for the cost of an unpainted dresser?

   A $32

   B $160

   C $320

   D $360

6. Thomas earns a 12.3% commission on the newspapers he sells each week. Last week, he sold $252 worth of newspapers. Which is the best estimate for the amount of money he earned in commission from last week's sales?

   A $25

   B $30

   C $208

   D $300

©ECS Learning Systems, Inc.

All rights reserved

Practice 2.B2

II.B Estimate solutions to application problems involving percent, including discounts, markups, commissions, and interest

1. Ashley wants to buy an aquarium from the pet store. The cost of the aquarium is $32.39, but Ashley will receive a 15% student discount. Which is the best estimate for the final price of the aquarium?

   A $48

   B $31

   C $27

   D $5

2. Each month, Abdul donates 17.5% of his allowance to charity. If he gave $12.85 to charity last month, which is the best estimate for the amount of allowance he received?

   A $7

   B $70

   C $95

   D $218

3. Ricki's pet snake laid 56 eggs. The pet store owner told her that about 28% of the eggs would **NOT** hatch. Which is the best estimate for the number of baby snakes Ricki should expect to hatch?

   A 52

   B 42

   C 18

   D 1.8

4. Lee completed an obstacle course in 14.87 minutes. It took his friend Tony 6% longer to finish the course. Which is the best estimate for the time it took Tony to finish the obstacle course?

   A 24 min

   B 21 min

   C 16 min

   D 0.16 min

5. Gail earned $67.20 in commission on the bouquets of flowers she sold in 3 days. If Gail sold $336 worth of flowers, which is the best estimate for the percentage of her commission?

   A 1.7%

   B 5.6%

   C 6%

   D 20%

6. There are 278 students at Ming's school. The principal announced that 62% of the students had grades good enough to make the honor roll. Which is the best estimate for the number of students that made the honor roll?

   A 48

   B 150

   C 180

   D 240

©ECS Learning Systems, Inc.

All rights reserved

Practice 2.B3

II.B Estimate solutions to application problems involving percent, including discounts, markups, commissions, and interest

1. Jeff returned a computer game that cost $56.19. The store charged him a 12% restocking fee. Which is the best estimate for the amount of money Jeff received after the restocking fee was deducted?

   A $5.50

   B $45

   C $50

   D $63

2. Mrs. Jameson must make a down payment of 21% of the cost of a new car. Which is the best estimate for the cost of the new car if the down payment is $3,856?

   A $20,000

   B $14,000

   C $3,000

   D $1,900

3. Each month, Freddy's Deli receives a 6% discount for every purchase of meat worth $4,000 or more. This month, the deli plans to purchase $4,168 of meat. Which is the best estimate for the discount the deli will receive?

   A $200

   B $240

   C $2,400

   D $4,000

4. At the end of the year, Mrs. Vasquez's employer pays her a 1% bonus on the net sales she made that year. This year, her net sales were $87,459. Which is the best estimate for the amount of Mrs. Vasquez's bonus check this year?

   A $87.50

   B $875

   C $1,000

   D $8,700

5. Jorge bought running shoes that cost $68. The sales tax in his community is 8.15%. Which is the best estimate for the total sales tax Jorge paid?

   A $5.60

   B $8.15

   C $56.00

   D $62.40

6. A sea turtle lays 74 eggs on a beach that is 12 miles long. Only 38% of the eggs actually hatch. Which is the best estimate for the number of baby turtles that will **NOT** hatch on the beach?

   A 4

   B 28

   C 30

   D 45

©ECS Learning Systems, Inc.

All rights reserved

Practice 2.B4

II.B Estimate solutions to application problems involving percent, including discounts, markups, commissions, and interest

1. Carver Middle School receives a 9.4% commission on all the candy the students sell during a fundraiser. Which is the best estimate for the amount of money the school will receive if the students sell $2,319.45 worth of candy?

   A $187

   B $230

   C $2,070

   D $2,300

2. The students ranked in the top 5% of their class earn a medal for excellence from the student council. There are 679 students in the seventh grade. Which is the best estimate for the number of seventh-grade students who will receive medals this year?

   A 350

   B 35

   C 30

   D 12

3. The band club sold $1,358 worth of carnations during a Valentine's Day fundraiser. They received $611 in commission from their sales. Which is the best estimate for the percentage of commission they earned?

   A 200%

   B 50%

   C 45%

   D 4.5%

4. Jamal purchased 4 packages of notebook paper for a total of $2.83. Including tax, the final cost of the paper was $3.05. Which is the best estimate for the percent of sales tax Jamal was charged on the paper purchase?

   A 1.08%

   B 1.1%

   C 6%

   D 8%

5. At the 10K fun run, 44 children completed the race in less than 1 hour. This represented 16% of the total number of children who ran in the fun run. Which is the best estimate for the total number of children who participated in the fun run?

   A 30

   B 250

   C 300

   D 450

6. Rex saved $15 to buy a gift for his mother. The gift he bought cost $12.05. Which is the best estimate for the percentage of his money Rex spent on the gift?

   A 80%

   B 70%

   C 50%

   D 8%

©ECS Learning Systems, Inc.

All rights reserved

Practice 2.B5

II.B Estimate solutions to application problems involving percent, including discounts, markups, commissions, and interest

1. Crystal earns a 19% commission on the clothes she sells at a department store. Which is the best estimate for the amount of commission she can expect to receive if she sells $853 worth of clothes this month?

A $160

B $170

C $425

D $1,600

2. Each month, Martin spends $14.67 of his $20 allowance at the movies. Which is the best estimate for the percentage of his allowance Martin spends on the movies each month?

A 7.5%

B 15%

C 25%

D 75%

3. Gretchen ran 5.23 miles last Saturday. This Saturday, she ran 21% farther. Which is the best estimate for how far Gretchen ran this Saturday?

A 1 mi

B 5.5 mi

C 6 mi

D 10.5 mi

4. Each month, Jocelyn earns 1.3% interest on the money in her college savings account. She currently has $5,412 in her college savings. Which is the best estimate for the amount of money she can expect to earn in interest this month?

A $6.50

B $70

C $500

D $650

5. Lucas fed his dog, Shy, 7 ounces of food each day. The veterinarian told Lucas to increase the amount of food he gave Shy to 9.5 ounces. Which is the best estimate for the percentage increase in food Lucas needs to start feeding Shy?

A 2%

B 20%

C 30%

D 75%

6. William receives $11.90 a week for mowing Mr. Wong's yard. He can earn an extra 36% each week if he trims the hedges. Which is the best estimate for the total amount of money William will receive if he trims the hedges and mows the lawn?

A $43

B $16

C $14

D $4.20

©ECS Learning Systems, Inc.
All rights reserved

Practice 2.B6

II.B Estimate solutions to application problems involving percent, including discounts, markups, commissions, and interest

1. A local store buys cookies for $1.98 a box and then sells them for $2.53 a box. Which is the best estimate for the percent increase in the price of 1 box of cookies?

   A 2.5%

   B 9.9%

   C 20%

   D 25%

2. Alyssa earns $25.25 a month for delivering groceries to the elderly. Of the money she earns, she spends $9.80 on bus fare each month. Which is the best estimate for the percentage of money Alyssa has left each month after paying her bus fares?

   A 65%

   B 60%

   C 35%

   D 15%

3. Last year, Betsy swam 27 laps in an endurance race. She increased the number of laps she swam by 37% in the last 6 months. Which is the best estimate for the number of laps Betsy can swim now?

   A 10

   B 33

   C 37

   D 65

4. Solomon's bank adds 2.1% interest to his checking account each month if his balance is over $3,000. This month, Solomon's checking balance was $3,825. Which is the best estimate for the amount of interest Solomon's account earned this month?

   A $60

   B $76

   C $800

   D $3,900

5. Dr. Sayegh usually charges $61.25 for a doctor's visit. With her insurance, Lea pays only $44.18 for each visit. Which is the best estimate for the percentage of each doctor's visit the insurance company pays for?

   A 13%

   B 25%

   C 66%

   D 75%

6. A giraffe increases its food intake by 32% in the winter, compared to the amount it eats in the summer. If a giraffe consumes an average of 79.4 pounds of food each day in the summer, which is the best estimate for the amount of food the giraffe will consume each day in the winter?

   A 104 lb

   B 94 lb

   C 86 lb

   D 56 lb

©ECS Learning Systems, Inc.

All rights reserved

Practice 2.C1

II.C Estimate solutions to application problems involving measurement

1. Manuel's empty pencil sharpener has enough room to hold 18.3 grams of pencil shavings. Manuel plans to sharpen all 54 of his map pencils. Which is the best estimate for the total number of times Manuel will empty the pencil sharpener if each pencil produces 1.69 grams of shavings?

   A 3

   B 4

   C 5

   D 6

2. Adrienne attached string to 8 kites. She attached 25.84 meters of string to 3 kites and 21.5 meters of string to 5 kites. Which is the best estimate for the total amount of string Adrienne attached to the kites?

   A 75 m

   B 175 m

   C 200 m

   D 300 m

3. Paula made lemonade for a lemonade stand. She could fill 18 small cups with 1 gallon of lemonade. If Paula sold 87 cups of lemonade over a 3-day period, which is the best estimate for the total number of gallons of lemonade Paula made in 3 days?

   A 1.8 gal

   B 4.2 gal

   C 5 gal

   D 5.5 gal

4. A small restaurant can serve 21 customers during the 1-hour lunch rush. Which is the best estimate for the amount of time it would take to serve each customer if there were only 1 server?

   A 3 min

   B 2.1 sec

   C 21 sec

   D 3 sec

5. James made 3 large sails and 22 small sails for a boat used in the school play. He used 9.75 meters of fabric for each large sail and 6.4 meters of fabric for each small sail. Which is the best estimate for the total amount of fabric he used to make all the sails?

   A 2.2 km

   B 218 m

   C 170 m

   D 140 m

6. Aaron wants to build 7 shelves that are each 38 inches long and 1 foot deep. He has a board that is 9 feet 8 inches long. Which is the best estimate for how much more wood he needs to build the shelves?

   A 15 ft

   B 14 ft

   C 13 ft

   D 10 ft

©ECS Learning Systems, Inc.

All rights reserved

Practice 2.C2

II.C Estimate solutions to application problems involving measurement

1. Henry used a vacuum cleaner that could hold up to 3.15 liters of debris. Which is the best estimate for the amount of debris Henry vacuumed up if he emptied the full vacuum 7 times?

   A 28 L

   B 21 L

   C 15 L

   D 10 L

2. The football team consumes 38.6 liters of sports drink during each game. There are 53 players on the team. Which is the best estimate for the average amount of sports drink each player consumes during a game?

   A 80 mL

   B 0.6 L

   C 0.8 L

   D 1.25 L

3. Horace is making an American flag to display on the football field during the half-time presentation. He purchased $34\frac{1}{2}$ yards of fabric to make 7 red stripes and $25\frac{1}{3}$ yards of fabric to make 6 white stripes. Which is the best estimate for the amount of fabric available for each red stripe?

   A 12 yd

   B 9 yd

   C 7 yd

   D 5 yd

4. A gallery displays photographs on a wall that is 6.12 meters wide. Each framed photograph is 38.5 centimeters wide. Which is the best estimate for the greatest number of framed photographs that could be placed side-by-side across the entire width of the gallery wall?

   A 15

   B 12

   C 6.7

   D 1.5

5. Sadie made 2 batches of cupcakes for a bake sale. Each batch produced 2 dozen cupcakes and used 256 milliliters of milk. Which is the best estimate for the amount of milk contained in each cupcake?

   A 20 mL

   B 15 mL

   C 10 mL

   D 5 mL

6. Charlize collected perfume samples. She had 4 samples that contained 0.38 ounces, 2.1 ounces, 0.76 ounces, and 0.9 ounces of perfume. Which is the best estimate for the total amount of perfume Charlize had?

   A 2.4 oz

   B 3.2 oz

   C 4 oz

   D 4.5 oz

©ECS Learning Systems, Inc. All rights reserved

Practice 2.C3

II.C Estimate solutions to application problems involving measurement

1. It took Jonah 1.85 minutes to finish reading 1 page of a novel. If there are 254 pages in the novel, which is the best estimate for the amount of time it will take Jonah to finish reading it?

   A 4 hr
   B 4 hr 45 min
   C 7 hr
   D 8 hr

2. An apartment manager painted 22 bedrooms in 1 building. He used 16.2 gallons of paint to finish the job. Which is the best estimate for the amount of paint it took to paint each bedroom?

   A 0.8 gal
   B 1 gal
   C 1.25 gal
   D 1.4 gal

3. A pottery company uses about 227 kilograms of clay each day to produce 528 clay pots. Which is the best estimate for the average weight of each clay pot?

   A 400 g
   B 2.5 kg
   C 5 kg
   D 40 kg

4. Mrs. Lowell covered 8 dining room chairs with new fabric that cost $3.98 per yard. Each chair required $2\frac{2}{3}$ yards of fabric. Which is the best estimate for the total amount of fabric needed for all 8 chairs?

   A 6 yd
   B 16 yd
   C 24 yd
   D 60 ft

5. Mr. Marsh spends $2\frac{3}{4}$ hours each day preparing meals in the school cafeteria. Which is the best estimate for the total amount of time he spends preparing meals in 3 months (60 school days)?

   A 20 hr
   B 180 hr
   C 540 hr
   D 570 hr

6. Mount Logan has an altitude of 5,959 meters. King Peak has an altitude of 5.17 kilometers. Which is the best estimate for the difference in the 2 altitudes?

   A 1 km
   B 100 m
   C 80 m
   D 10 km

©ECS Learning Systems, Inc.

All rights reserved

Practice 2.C4

II.C ***Estimate solutions to application problems involving measurement***

1. It takes 8.3 minutes to fill a 50-gallon barrel with water from the garden hose. Which is the best estimate for the amount of time it will take to fill 7 barrels that each hold 25 gallons of water?

   A 16 min
   B 20 min
   C 28 min
   D 35 min

2. Lynda made a frame for her painting. She used 2 rectangular pieces of wood that were each 42.5 centimeters long and 2 pieces that were each 54.25 centimeters long for the sides. Which is the best estimate for the perimeter of the frame?

   A 90 cm
   B 170 cm
   C 200 m
   D 2 m

3. Clifton learned that the average 60-watt light bulb will last about 450 hours before burning out. His light bulb has been burning constantly for 11 days. Which is the best estimate for how much longer his light bulb will last before burning out?

   A 200 hr
   B 440 hr
   C 20 days
   D 10 days

4. A bean plant grows at a steady rate of 4.8 millimeters per day from the time it sprouts. Which is the best estimate for the number of days it would take the bean plant to reach a height of 6.25 centimeters?

   A 1.3 days
   B 6.5 days
   C 10 days
   D 13 days

5. Candice used 9 oranges to make 262 milliliters of freshly squeezed orange juice. Which is the best estimate for the number of oranges she would need to make 4 liters of juice?

   A 15
   B 65
   C 150
   D 650

6. In 1 week, Layton drove his go-cart around a circular track 53 times.

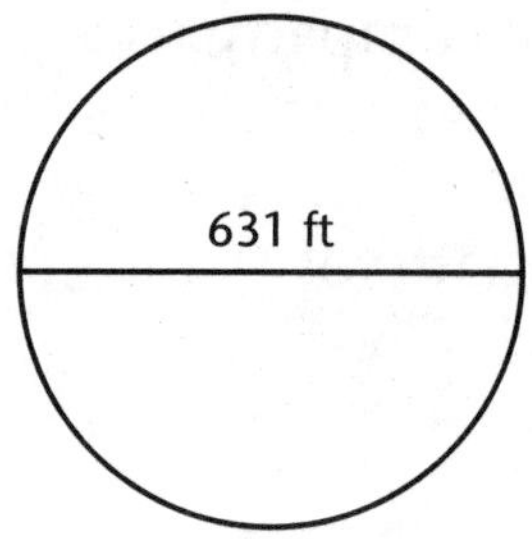

   Using the given information, which is the best estimate for the distance Layton drove his go-cart in 1 week? (Use 3 for π)

   A 10 mi
   B 20 mi
   C 30 mi
   D 40 mi

©ECS Learning Systems, Inc.

All rights reserved

Practice 2.C5

II.C Estimate solutions to application problems involving measurement

1. Garrett's computer printer can print 4.75 pages per minute. Which is the best estimate for the time it would take the printer to produce a 94-page document?

   A 50 min

   B 25 min

   C 20 min

   D 5 min

2. One marble weighs 4.9 grams. Which is the best estimate for how many marbles are in a pouch weighing 1.5 kilograms?

   A 300

   B 250

   C 30

   D 3

3. Mark's gecko grows at a rate of 6.3 millimeters per day from the time it hatches. Which is the best estimate for the length Mark's gecko will be after 6 weeks (42 days)?

   A 36 mm

   B 60 mm

   C 240 mm

   D 2.4 m

4. Kareem plays basketball for his school. During practice, he ran the length of the court 28.5 times before his coach blew the whistle. Which is the best estimate for the distance Kareem ran during practice?

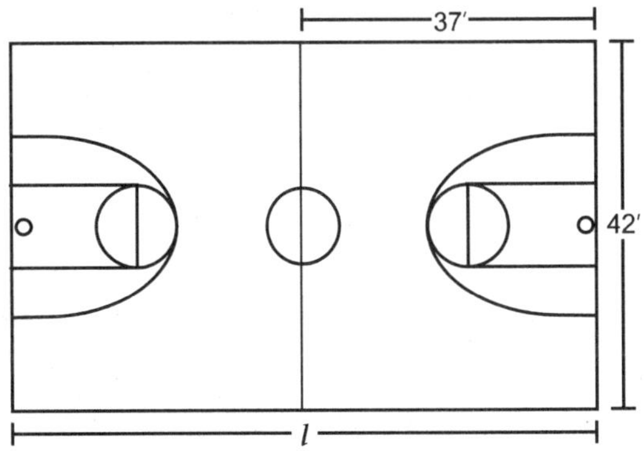

   A 1,000 ft

   B 1,200 ft

   C 1,800 ft

   D 2,400 ft

5. Madison used the following ingredients for a punch recipe: 0.61 liters of soda, 500 milliliters of raspberry tea, and 0.39 liters of pineapple juice. Which is the best estimate for the amount of punch the recipe made?

   A 15 L

   B 10.5 L

   C 1.5 L

   D 150 mL

©ECS Learning Systems, Inc.

All rights reserved

Practice 2.C6

II.C Estimate solutions to application problems involving measurement

1. Mario runs at a rate of 9.7 kilometers per hour. Which is the best estimate for the length of time it will take him to run 5 kilometers?

   A 2 hr

   B 1.6 hr

   C 40 min

   D 30 min

2. One piece of gum weighs 3.61 grams. Which is the best estimate for the number of pieces of gum in a bag weighing 1.42 kilograms?

   A 400

   B 300

   C 42

   D 38

3. Jolene can jump rope at a rate of 48 jumps per minute. Which is the best estimate for the number of times she can jump in 7 minutes?

   A 550

   B 400

   C 350

   D 240

4. An elevator rises at a rate of 1 floor every 4.8 seconds. Each floor is 9 feet tall. Which is the best estimate for the amount of time it will take the elevator to travel 243 feet?

   A 1.5 min

   B 2 min

   C 4 min

   D 5 min

5. A candle burns down at a rate of 12.5 millimeters per hour. Which is the best estimate for the amount of time it will take a candle that is 21.5 centimeters tall to burn completely?

   A 15 min

   B 0.6 hr

   C 1.5 hr

   D 15 hr

6. Savannah wants to cut 24-inch logs from a 47.2-foot tree trunk. Which is the best estimate for the number of logs she can cut from the tree trunk?

   A 24

   B 18

   C 10

   D 2

©ECS Learning Systems, Inc.

All rights reserved

Notes

©ECS Learning Systems, Inc.

All rights reserved

Solution Strategies/ Analyze or Solve Problems

III. Determine solution strategies and analyze or solve problems.

A. Solve application problems involving ratios and proportions, including scaling and unit costs

B. Solve application problems involving percent, including discounts, markups, commissions, and interest

C. Identify missing or extraneous information in application problems

©ECS Learning Systems, Inc. All rights reserved

Objective 3: Pretest

III.A Solve application problems involving ratios and proportions, including scaling and unit costs (1–8)

1. A newspaper ad shows a scale drawing of a house. In the drawing, 2 inches represent 10 feet in the actual house. If the master bedroom in the house is actually 15 feet long, how long is it in the scale drawing?

   A 2.5 in

   B 3 in

   C 3.5 in

   D 5 in

2. A book includes a map drawn to scale, with 3 centimeters = 75 miles. On the map, a river is 7 centimeters long. How long is the actual river?

   A 525 mi

   B 225 mi

   C 175 mi

   D 150 mi

3. At a garden shop, Bonnie bought a 15-pound bag of sand for $7.50. She used 8 ounces of sand in each of her flower boxes at home. How much did Bonnie pay for the sand used in 1 flower box?

   A $1

   B 75¢

   C 50¢

   D 25¢

4. Mrs. Sandberg deducts 3 points for every 5 misspelled words on a student's paper. How many points would be deducted from a paper that contained 12 misspelled words?

   A 7.2

   B 8.4

   C 9

   D 20

5. A store sells colored pencils in sets of 12 for $6.90. The store also sells single pencils for the unit price. How much would you pay for 4 single pencils?

   A $3.00

   B $2.75

   C $2.30

   D $2.20

6. A machine can process mail at a rate of 200 pieces every 40 seconds. At this rate, how many pieces of mail can the machine process in 4 minutes?

   A 800

   B 960

   C 1,000

   D 1,200

7. Morgan followed a set of printed directions to make a necklace. For every 2.5 inches shown on the necklace pattern, she used 14 beads. How many beads did Morgan use if the pattern showed a total of 15 inches?

   A 70

   B 84

   C 112

   D 210

©ECS Learning Systems, Inc. All rights reserved

8. In art class, Jake drew a picture of the school playground. He drew the picture to scale, with 3 centimeters equal to 5 feet. If the tallest slide on the playground is 9 feet tall, how tall would it have been in Jake's drawing?

A 15 cm

B 5.6 cm

C 5.4 cm

D 1.66 cm

III.B Solve application problems involving percent, including discounts, markups, commissions, and interest (9–16)

9. A store announced a 3.5% increase in prices beginning February 1. A coat cost $42.99 in January. Rounded to the nearest dime, how much would the same coat cost on February 1?

A $44.50

B $46.50

C $47.50

D $58.00

10. A total of 1,240 students attend classes at Martí Middle School, but only 62 students are members of the track team. What percentage of students are members of the track team?

A 5%

B 5.4%

C 6%

D 20%

11. Maurice borrowed $25 from his brother, who charged him 5% interest on the loan. When Maurice repays the loan, he will owe his brother a total of—

A $25.05

B $25.50

C $26.25

D $26.35

12. A store owner can buy a case of soft drinks for $3.60. Each case contains 24 cans of soda. Customers pay 45¢ for 1 can of soda. What percentage is the markup for each can of soda?

A 3%

B 30%

C 33%

D 300%

13. Mrs. Lorenzo sells used cars and earns a 5.5% commission on her total sales each month. In May, her sales totaled $24,000. What commission did Mrs. Lorenzo earn in May?

A $1,120

B $1,200

C $1,320

D $1,350

14. Last year, Mr. Kim paid a total of $1,235 for electricity. This year he wants to reduce the electricity bill by 15%. If he reaches his goal, how much will Mr. Kim pay for electricity this year?

A $1,220.00

B $1,049.75

C $1,019.55

D $185.25

©ECS Learning Systems, Inc.
All rights reserved

15. Jenny's grade on her science project was an 84. Jenny's teacher increased her grade by 5% because she turned the project in early. How many points did the teacher add to Jenny's grade?

A 4

B 4.2

C 4.5

D 5

16. At a bookstore, Glenna bought a Reader One Card for $5. She can use the card to receive a 12% discount on her purchases at the store. Last week, Glenna bought books that cost a total of $65 before the discount. How much did Glenna save on the books she bought?

A $5.00

B $7.70

C $7.80

D $12.00

III.C Identify missing or extraneous information in application problems (17–24)

17. Cassandra spent 50% of her study time on her math project and 35% of her study time on her science project. What other information is needed to determine how much time Cassandra spent on her math project?

A the total amount of time she spent studying

B what she did the other 15% of the time

C the number of people who worked with her on the math project

D the math project's deadline

18. Stephen needs 13 meters of wood to build a dresser for his customer. He purchased 5 boards that were 3 meters long and spent $4.50 on each board. Which of the following questions **cannot** be answered using the given information?

A How much money did Stephen spend on the boards in all?

B What was the total length of all of the boards?

C How long was the dresser?

D How many 3-meter boards were needed to build the dresser?

19. Mrs. Henderson gave her baby 5 ounces of formula regularly during the daytime. She also fed the baby 7 ounces of peas, carrots, or yams each day. What information is necessary to determine how much formula Mrs. Henderson gave her baby each day?

A the amount of time the baby slept during naptime

B the number of times she fed the baby peas, carrots, or yams

C the length of the daytime

D the total number of times the baby drank formula each day

20. Shelly spends $45.60 on 14 souvenirs at the beach. She divides the souvenirs equally among her friends at home. What information is needed to determine the cost of each friend's souvenirs?

A the amount of time Shelly spent at the beach

B the number of friends Shelly has

C the total cost of the beach trip

D the average cost of each souvenir

©ECS Learning Systems, Inc. All rights reserved

21. Mr. White estimated that it would cost $450 to build a fence for a customer. After the 650-yard fence was built, he wanted to determine how much the job actually cost. He knew that 3 employees worked on the job for 8 days. He also knew that the supplies cost $45 for every 38 yards of the fence. Which of the following questions **cannot** be answered using the given information?

A How much did the supplies cost in all?

B How much money did each employee earn every day?

C How much did the fence supplies cost per yard?

D What was the company's original estimate for the job?

22. Wanting to spend no more than an average of $120 per room, Mary Anne decided to redecorate her house. She spent $235 on new furniture, $159 on new rugs, and $56 on new paint. She also paid a company $300 to do all of the wallpapering. What other information is necessary to determine if Mary Anne spent more money than she originally intended?

A the number of square feet of the painted walls

B the number of rooms with new furniture

C the number of rooms that needed wallpaper

D the number of rooms she redecorated

23. Dr. Peche has a total of 1,349 patients on file. Each day, he spends 5 hours at his downtown office, 1 hour at lunch, and the rest of his 8-hour workday at his north-side office. Additionally, he works an average of 12 hours per week at the hospital. Which of the following questions **cannot** be answered using the given information?

A On average, how many total hours does Dr. Peche work each week?

B How many hours does Dr. Peche work downtown each week?

C What is the average amount of time Dr. Peche spends with each patient?

D On average, how much time does Dr. Peche spend at the hospital each year?

24. Terry knows that 4 gallons of juice cost $4.50. She also knows that it takes 2 gallons of juice to make 5 gallons of punch. What else does Terry need to know to determine how much it will cost to buy enough juice to make punch for a birthday party?

A the cost of the other punch ingredients

B the amount of punch she must make at the party

C the amount of ice required for the punch

D the price Terry's catering service charges per cup of punch

©ECS Learning Systems, Inc. All rights reserved

Practice 3.A1

III.A Solve application problems involving ratios and proportions, including scaling and unit costs

1. Mr. Coker spends $40 of his $560 paycheck on gas each month. How much money has he spent on gas after making $8,960?

   A $640

   B $636

   C $224

   D $25

2. Sarah, a freelance writer, wrote a 450-word essay and was paid $112.50. She wrote a 525-word essay based on the same pay scale. How much was Sarah paid for the 525-word essay?

   A $96.43

   B $130.67

   C $131.23

   D $131.25

3. Carolyn paints a picture of a house using the scale 2 centimeters = 1.6 feet. The house's front window is 5.6 feet long. How long is it in the painting?

   A 0.57 ft

   B 0.14 cm

   C 6 cm

   D 7 cm

4. Trevor pays $12 for an 0.8-ounce tube of acrylic paint. How much does the paint cost per ounce?

   A $450

   B $150

   C $16

   D $15

5. A gas station charged one customer $15 for 12 gallons of gasoline. How much did the station charge another customer who purchased 9 gallons of the same kind of gasoline?

   A $7.20

   B $11.25

   C $11.30

   D $20

6. Lennox built dollhouse furniture based on a scale of 0.5 inches = 3 feet. He made a model of an actual bed that was 5.1 feet wide. What was the width of the dollhouse bed?

   A 0.65 in

   B 0.85 in

   C 8.5 in

   D 30.6 in

©ECS Learning Systems, Inc.

All rights reserved

Practice 3.A2

III.A Solve application problems involving ratios and proportions, including scaling and unit costs

1. Mrs. Drifet buys special art markers in packages of 42. Each package costs $18.90. How much does each marker cost?

   A 44¢

   B 45¢

   C $2.20

   D $4.50

2. Sheena must buy concert tickets for the radio station where she works. She is willing to pay $312 for 48 tickets. If she decides to pay $416 instead, how many tickets should she receive if she is paying the same price per ticket?

   A 36

   B 48

   C 64

   D 2,704

3. Mr. Tavya spends $120 on 3 bales of hay for his horses. How much money does he spend if he purchases 17 bales of hay?

   A $21.18

   B $660

   C $680

   D $2,040

4. Theresa spent $51 to purchase 30 heartworm pills for her dog. If the dog must take 1 pill each day, how much money does Theresa spend on the dog's pills each week (7 days)?

   A $11.90

   B $11.27

   C $4.12

   D 59¢

5. Jeremiah builds model cars. One model's hood is 6 inches wide, and the trunk is 5.25 inches wide. If the hood is 4 feet wide on the actual car, how wide is the actual trunk?

   A 3.3 ft

   B 3.5 ft

   C 4.58 ft

   D 7.88 ft

6. Raquel can paint a wall that is 52 square feet using 0.7 gallons of paint. How much paint will she use to paint several walls that have a total area of 442 square feet?

   A 5.95 gal

   B 12.14 gal

   C 59.5 gal

   D 309.4 gal

©ECS Learning Systems, Inc. All rights reserved

Practice 3.A3

III.A Solve application problems involving ratios and proportions, including scaling and unit costs

1. Each week, Morgan spends 3 hours out of 25 volunteer hours reading to children at a hospital. How many hours does she spend reading to the children each month (4 weeks)?

A 0.48 hr

B 1.3 hr

C 12 hr

D 33 hr

2. Dominic earned 5 B's for every 2 A's in his math class during one six weeks. If he earned a total of 35 B's, how many A's did he earn?

A 12

B 14

C 15

D 87.5

3. Good Stuff Grocery sells 3 spears of broccoli for $1.20. If Sydney buys 7 spears of broccoli, how much money does she spend?

A $17.50

B $8.20

C $2.80

D $2.24

4. A radio station talk show has 3 minutes of news updates for every 0.5 minutes of commercials each morning. If there are 17 minutes of commercials one morning, how many minutes of news updates are there?

A 2.83 min

B 11.3 min

C 25.5 min

D 102 min

5. The grocery store clerks at A-1 Groceries must replace their cash drawers twice every 5.5 hours. If one checker replaced his cash drawer 16 times in 1 week, how many hours did he work?

A 4.4 hr

B 5.8 hr

C 44 hr

D 88 hr

6. Sierra paid $45.60 for her long-distance phone bill last month. Her phone company charged 12¢ per minute for each call. This month, Sierra got a discount rate of 9¢ per minute for each long-distance call. If Sierra spends the same amount of time on the phone this month as last month, how much will her long-distance phone bill be?

A $34.20

B $35.20

C $43.20

D $60.80

©ECS Learning Systems, Inc.

All rights reserved

Practice 3.A4

III.A Solve application problems involving ratios and proportions, including scaling and unit costs

1. Samantha bakes $2\frac{1}{2}$ batches of cookies using a total of 6 cups of sugar. If she makes $5\frac{1}{2}$ batches of cookies, how many cups of sugar will she use?

   A 6 c

   B 9 c

   C $13\frac{1}{5}$ c

   D 18 c

2. It takes Paul 5 minutes to clean 9 panels of glass on a French door. How long will it take him to clean 45 panels of glass?

   A 9 min

   B 25 min

   C 26 min

   D 81 min

3. Natalie spends 11.6 hours in 2 weeks practicing for her karate competition. If her competition is still 5 weeks away, and she practices at the same rate each week, how many more hours will she practice before the competition?

   A 11 hr

   B 27.5 hr

   C 29 hr

   D 58 hr

4. At the pier, a fisherman sells 0.75 pounds of tuna for $1.35. How much will it cost if Mackenzie wants to buy 3.25 pounds of tuna from the fisherman?

   A $5.85

   B $5.75

   C $3.21

   D $1.81

5. The diagram below is a scale-drawing of Cameron's rectangular bedroom. What is the actual length of his bedroom?

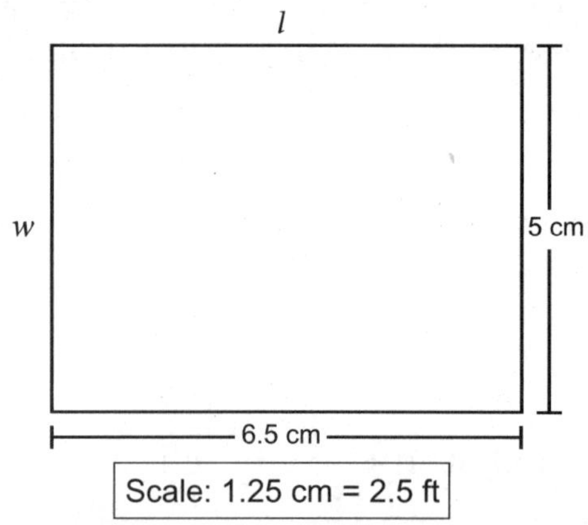

   A 13 ft

   B 12.1 ft

   C 10 ft

   D 3.25 ft

6. Angel has paid $3,492 on his car loan by making 6 equal monthly payments. If he has $5,238 left to pay on the loan, and he continues to make the same monthly payments, how many more months will it take him to pay off the loan?

   A 4

   B 8

   C 9

   D 10

©ECS Learning Systems, Inc. All rights reserved

Practice 3.A5

III.A Solve application problems involving ratios and proportions, including scaling and unit costs

1. Maria enlarged a rectangular photograph that was 3 inches wide by 6 inches long to fit into a rectangular frame that was 7 inches wide. How long was the frame?

   A 3.5 in

   B 10 in

   C 13 in

   D 14 in

2. Eyeglass Enterprises usually serves 78 customers and makes $1,404 each day. If they serve 130 people on Monday, how much money can they expect to make?

   A $2,460

   B $2,340

   C $234

   D $7.22

3. Carson's mashed potato recipe calls for $4\frac{1}{2}$ tablespoons of butter for every 5 potatoes used. How many tablespoons of butter will he need if he uses 12 potatoes?

   A $1\frac{7}{8}$ T

   B $10\frac{1}{5}$ T

   C $10\frac{4}{5}$ T

   D 11 T

4. A scale model of Jarod's rectangular garden is shown below. If the actual length of his garden is 7.2 feet, what is the actual width of the garden?

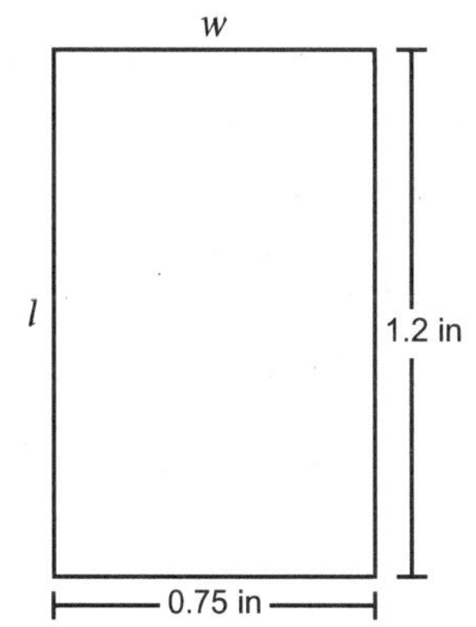

   A 45 ft

   B 15 ft

   C 11.52 ft

   D 4.5 ft

5. Kelsey built model airplanes using the same scale for each model. One model had a wingspan of 6.3 centimeters to represent an actual wingspan of 25.2 feet. What is the actual wingspan of a plane whose model has a 9.7-centimeter wingspan?

   A 244.44 ft

   B 38.8 ft

   C 28.6 ft

   D 2.43 ft

6. Cost Less Clothing sells 4 pairs of socks for $16.88. How much would it cost to buy 11 pairs of the same socks?

   A $2.61

   B $4.64

   C $46.42

   D $48.42

©ECS Learning Systems, Inc. All rights reserved

Practice 3.B1

III.B Solve application problems involving percent, including discounts, markups, commissions, and interest

1. Jada wants to buy a car worth $9,670. If she waits until the car goes on sale for 7% off, how much will it cost?

A $676.90

B $7,069

C $8,986.20

D $8,993.10

2. Monica makes 2% commission on every bridal gown she sells. If Monica sells $4,560 worth of gowns, how much is her commission?

A $90.20

B $91.20

C $912.00

D $4,468.80

3. Antoine buys a guitar worth $345 on sale for $276. What percent was his discount?

A 80%

B 20%

C 5%

D 1.25%

4. Stickers cost 8¢ each. If Martin buys 30 stickers, he gets a 10% discount. How much will Martin owe if he purchases 30 stickers?

A $2.16

B $2.40

C $21.60

D $24

5. Autumn makes $12 per hour. If she gets a 7.5% raise, how much will her hourly wage be?

A $21

B $12.90

C $11.10

D $9

6. Jason borrows $350 from his friend and promises to pay him 25% interest on the loan. How much money does Jason owe his friend, including interest?

A $87.50

B $358.75

C $437.50

D $875

©ECS Learning Systems, Inc. All rights reserved

Practice 3.B2

III.B Solve application problems involving percent, including discounts, markups, commissions, and interest

1. Jayme bought a $60 swimsuit on sale for $24. What percent discount did Jayme get?

   A 25%

   B 36%

   C 40%

   D 60%

2. Janet's Coffee Shop decided to increase prices by 12%. If a regular cup of coffee used to cost 75¢, how much will it cost after the price increase?

   A 9¢

   B 84¢

   C 90¢

   D $1.65

3. At his first job, Julian earned $3,560 each month. At his second job, he received an 11% increase in his monthly wage. What was Julian's monthly pay at his second job?

   A $391.60

   B $3,168.40

   C $3,916

   D $3,951.60

4. The Garfield family wanted to add on to their 1,450 square-foot house. If they increased their house size by 45%, how many square feet did they add?

   A 65.25 ft^2

   B 652.5 ft^2

   C 797.5 ft^2

   D 2,102.5 ft^2

5. An airplane flew at an altitude of 340 feet 30 seconds after taking off. One minute later, the plane flew at an altitude of 799 feet. By what percent did the plane's altitude increase in 1 minute?

   A 235%

   B 135%

   C 42.5%

   D 2.35%

6. The bank charges Mr. Zilker 9% interest on a $525 loan. How much must Mr. Zilker pay the bank in interest on his loan?

   A $572.25

   B $472.50

   C $47.25

   D $42.75

©ECS Learning Systems, Inc. All rights reserved

Practice 3.B3

***III.B* Solve application problems involving percent, including discounts, markups, commissions, and interest**

1. Kathy received a 15% discount on a bedspread that originally cost $180. How much did the bedspread cost after the discount?

A $27

B $153

C $163

D $270

2. A company's stock was worth $12.35 a share on Monday. On Friday, the stock was worth $14.82 a share. By what percent did the company's stock increase in value?

A 1.2%

B 12%

C 20%

D 83%

3. Mrs. Alvarado used a 40%-off coupon to pay for a bathroom cleanser that originally cost $5.65. How much did Mrs. Alvarado pay for the cleanser before tax?

A $2.24

B $2.26

C $3.39

D $5.42

4. A professional basketball player made 130 rebounds in his first season. The next season, he made 80% more rebounds than he did in his first season. How many rebounds did the basketball player make in his second season?

A 104

B 134

C 140

D 234

5. The average annual rainfall in Ottawa is 870 millimeters. The average annual rainfall in Vancouver is 168% of the average in Ottawa. What is the average annual rainfall in Vancouver?

A 2,331.6 mm

B 1,461.6 mm

C 1,460.6 mm

D 146.16 mm

6. Mr. Schnapp received a 14% commission for every sale he made at Electronics, Etc. If he sold $35,670 in electronic merchandise, what was his commission?

A $499.38

B $4,939.80

C $4,993.80

D $5,033.80

©ECS Learning Systems, Inc. All rights reserved

Practice 3.B4

III.B Solve application problems involving percent, including discounts, markups, commissions, and interest

1. Mrs. Stower's rent was $640 each month. When she renewed her lease, her rent increased by 7%. How much was her rent then?

A $1,088
B $684.80
C $448
D $44.80

2. Rachel received a 33% discount on a $52 sweater. How much did the sweater cost on sale?

A $17.16
B $34.84
C $44.84
D $45.16

3. The end of a soccer practice field is 75 meters long. The goal posts mark off 7.32 meters of the length of the end of the field. What percent of the end-line is marked off by goal posts?

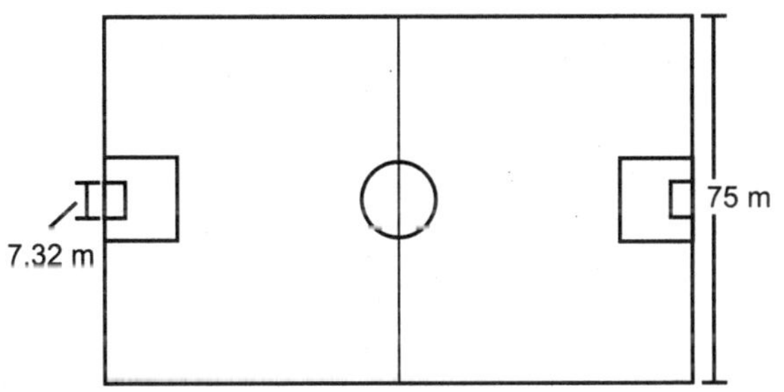

A 97.6%
B 90.2%
C 9.76%
D 0.09%

4. An investor put $14,509 into the stock market. She lost 56% of her investment. How much money did the investor lose?

A $6,383.96
B $8,025.04
C $8,124.54
D $8,125.04

5. Mrs. Bangle loaned her friend $45. When her friend repaid her, she added 15% interest. How much did Mrs. Bangle's friend repay her?

A $6.75
B $41.75
C $51.75
D $67.50

6. Sheila works for an amusement park as a cartoon character. She is not allowed to wear her costume for more than 35% of the time at work. If Sheila works 52 hours in 1 week, what is the maximum amount of time she can be in costume?

A 16.7 hr
B 17.2 hr
C 18.1 hr
D 18.2 hr

©ECS Learning Systems, Inc. All rights reserved

Practice 3.B5

III.B Solve application problems involving percent, including discounts, markups, commissions, and interest

1. The Mississippi River is 32.5% shorter than the Nile River. If the Nile River is 3,470 miles long, how long is the Mississippi River?

A 4,597.75 mi

B 2,342.25 mi

C 2,324.25 mi

D 1,127.75 mi

2. Leigh purchased a sweater for $48.75 plus 6.5% sales tax. How much was her total purchase (rounded to the nearest penny)?

A $51.92

B $51.82

C $31.69

D $3.17

3. In 1 year, Claire saved $1,246. At the end of the year, the bank added 2.5% interest to her savings account. How much money did Claire have in her savings account then?

A $31.15

B $1,276.05

C $1,277.15

D $1,557.50

4. During the summer, Kenneth earned $79.36. He spent $19.84 of his earnings on a new basketball. What percent of his money did Kenneth spend on the basketball?

A 2.5%

B 4%

C 25%

D 60%

5. A computer game originally cost $29.95. Victor bought it on sale for 20% off. How much did Victor pay for the computer game before tax?

A $5.99

B $23.96

C $24.16

D $35.94

6. Sean's Skateboards made a profit of $16,723 in May. If the profit in June was 45% larger than in May, what was the total amount of June's profit?

A $7,525.35

B $9,197.65

C $23,248.35

D $24,248.35

©ECS Learning Systems, Inc. All rights reserved

Practice 3.C1

III.C Identify missing or extraneous information in application problems

1. Phillipe spent 5 months training for a marathon. When his training began, he could run 12 miles in 2 hours 30 minutes. At the end of the 5 months, he could run 24 miles in 3 hours 37 minutes. At the actual marathon, he finished the 26-mile race in third place. Which of the following questions **cannot** be answered using the given information?

   A Approximately how long did it take Phillipe to run the marathon?

   B On average, how long did it take Phillipe to run 10 miles at the beginning of his training?

   C On average, how long did it take Phillipe to run 12 miles at the end of his training?

   D How many miles did Phillipe run during the 5-month training period?

2. Amber budgeted $345 each month (4 weeks) for food and gas purchases. She spent the following amount of money on food in the last 8 weeks: $65.67, $78.99, $54.23, $29.44. Which of the following pieces of information does Amber need to determine whether or not she was within her budget?

   A the percent of $345 designated for food

   B the amount of money she spent on food the previous 4 months

   C the amount of money she spent on gas in the last 8 weeks

   D the percent of her paycheck designated for food and gas

3. Trent and Lisa like to read the same books at the same time. Lisa reads an average of 45 pages each day. She usually reads 10 of those pages aloud. Trent reads an average of 52 pages each day and only reads 5 of those pages aloud. Which of the following pieces of information is necessary to determine the average number of books both Lisa and Trent read each month?

   A the length of time it takes Lisa to read 45 pages

   B the length of time it takes Trent to read 5 pages aloud

   C the page-length for 1 book Trent and Lisa read

   D the average number of pages per book that Trent and Lisa read

4. Mrs. Sartor made a chart showing the number of students she teaches each period of the day. She does not teach the same student twice in the same day.

Period	Number of Students
1	32
2	28
3	26
4	31
5	27

   Using the given information, you **cannot** determine—

   A the total number of students Mrs. Sartor teaches

   B Mrs. Sartor's average class size

   C which class size Mrs. Sartor prefers

   D Mrs. Sartor's median class size

©ECS Learning Systems, Inc. All rights reserved

Practice 3.C2

III.C Identify missing or extraneous information in application problems

1. One weekend, Jerry caught 12 butterflies and 10 fireflies. The next weekend, he caught 24 butterflies and 3 house flies. The third weekend, he caught 48 butterflies. If he continues to catch butterflies each weekend at the same rate for 10 weeks, which question **cannot** be answered using the given information?

   A How many butterflies did Jerry catch during the sixth weekend?

   B How many butterflies did Jerry catch in all during the 10 weeks?

   C How many butterflies can Jerry catch per hour?

   D How many more butterflies than other insects did Jerry catch the first 2 weekends?

2. Jane paints signs for businesses. Today, she must paint a white border around the perimeter of 17 different signs. The width of the border on each sign is 3 inches. She uses 3 kinds of paintbrushes and 1.67 gallons of paint to cover all of the perimeters. Which of the following pieces of information is necessary to determine how much paint Jane uses on each sign's border?

   A the number of sides on each sign

   B the total number of paintbrushes she uses

   C the amount of time it takes to paint each border

   D the measure of each sign's perimeter

3. Dexter set up his monthly budget on the chart below.

Food	Rent	Car	Utilities	Misc.	Savings
10%	40%	20%	15%	5%	

   Which of the following information is needed to determine the amount of money Dexter puts in savings each month?

   A the percent of money he dedicates to savings

   B the amount of money he earns each month

   C the amount of money he spends on gas each month

   D the balance of his savings account

4. Benjamin bought a piece of wood that was 5 feet long and 1 foot wide for \$3 off the original price in order to build a cutting board. The rectangular cutting board will be $2\frac{1}{2}$ feet long and 2 feet wide. Which of the following questions **cannot** be answered using the given information?

   A Does Benjamin have enough wood to make the cutting board?

   B What will the area of the cutting board be?

   C What is the perimeter of the original piece of wood?

   D How much is the cutting board worth?

©ECS Learning Systems, Inc.

All rights reserved

Practice 3.C3

III.C Identify missing or extraneous information in application problems

1. Mr. Piper moved his office from a building with 8,340 square feet to a building with 4,490 square feet. He had to sell $5,340 worth of business furniture to fit into the new office space. Rent for the new building was lower, but Mr. Piper still paid the same amount per square foot of space. Which of the following information is necessary to determine how much rent Mr. Piper paid for the new building?

   A the value of the office furniture

   B the number of offices in the new building

   C the rent for the old building

   D the percentage difference in the building sizes

2. An architect designed 3 houses; each house had 2,498 square feet of space. The floor plans for each house cost $560. It cost $25 per square foot to build each house. Which of the following information is needed to determine the amount of money the architect made by selling floor plans?

   A the cost of the floor plans per square foot

   B the number of floor plans the architect sold

   C the amount the architect charged to design a house

   D the number of houses built

3. During the first 2 years she owned her car, Kaori drove 25,943 miles and spent the following amounts on maintenance.

oil changes (x 8)	$173
check-ups (x 3)	$139
other	$267

   The car dealership reimbursed Kaori for all maintenance that occurred during the car's first 2,000 miles. Which of the following questions **cannot** be answered using the given information?

   A What was the total amount of money Kaori spent on her car in the first 2 years?

   B How much money did the dealership reimburse to Kaori?

   C What is the average number of miles Kaori put on her car each month?

   D What was the average cost of each oil change?

4. Aubree cut, colored, and styled Kay's hair. Kay's total cost, including 5% tax, was $59.33, 60% of which was for the haircut. Which of the following questions **cannot** be answered using the given information?

   A How much did Aubree charge to color Kay's hair?

   B How much did Aubree charge Kay for just the haircut?

   C How much did Kay pay in tax?

   D What did Kay pay for Aubree to just color and style her hair?

©ECS Learning Systems, Inc.

All rights reserved

Practice 3.C4

III.C Identify missing or extraneous information in application problems

1. Marta's Flower Shop sells a total of \$456,987 worth of flowers in 1 year. Of the year's total sales, 38% occur in the months of February, May, and December. Which of the following questions **cannot** be answered using the given information?

   A How much money does the shop make in December?

   B How much money does the shop make in February, May, and December?

   C How much money does the shop make during the remaining 9 months of the year?

   D What is the average amount of money the shop makes each month?

2. Shalom has a socket set with 14 sizes of sockets, a socket wrench, and 3 wrench attachments. The entire set is worth \$45, but Shalom paid only \$35 for it. Which of the following questions **cannot** be answered using the given information?

   A What is the average cost of each piece of the socket set?

   B What is the ratio of wrench attachments to sockets?

   C What is the average cost of each individual socket?

   D What percent discount did Shalom get on the socket set?

3. The pie graph shows the percentage of people of different ages who purchased a new cookie product at the grocery store.

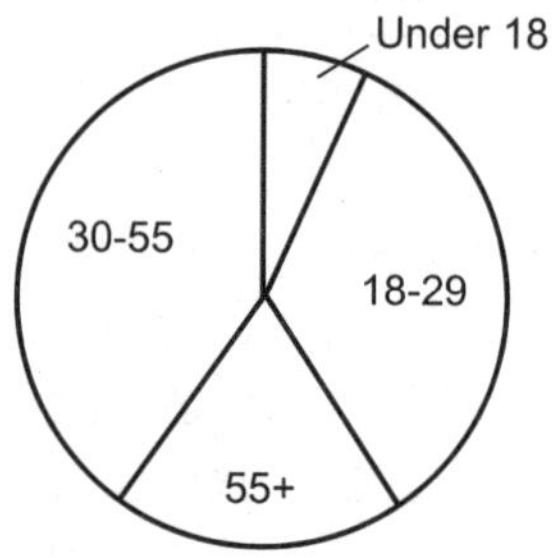

   Which of the following questions can be answered using the given information?

   A What percent of people ages 18–55 purchased the cookie product?

   B What is the ratio of people ages 30–55 to 18–29 who purchased the product?

   C How many people under the age of 18 purchased the product?

   D Did more people 30 and over or younger than 30 purchase the product?

4. Mara has 2 pairs of shoes for every 3 outfits she owns. All of Mara's shoes are worth \$156. Which of the following information is necessary to determine the number of outfits Mara owns?

   A the amount of money all of Mara's outfits are worth

   B the number of pairs of shoes Mara owns

   C the average cost of each outfit

   D how many more shirts than pants Mara owns

©ECS Learning Systems, Inc.

All rights reserved

Practice 3.C5

III.C Identify missing or extraneous information in application problems

1. Mr. Dole must refill the gas tank of his riding lawn mower every 150 rows he mows. He rides a distance of 3 miles to mow his lawn completely. Which of the following information is necessary to determine how often Mr. Dole must refill his lawn mower's gas tank if he wants to mow the entire yard?

A the length of each row

B the area of his yard

C the size of the gas tank on the lawn mower

D the amount of time it takes him to mow the lawn completely

2. The diagram below is a model of Amanda's room.

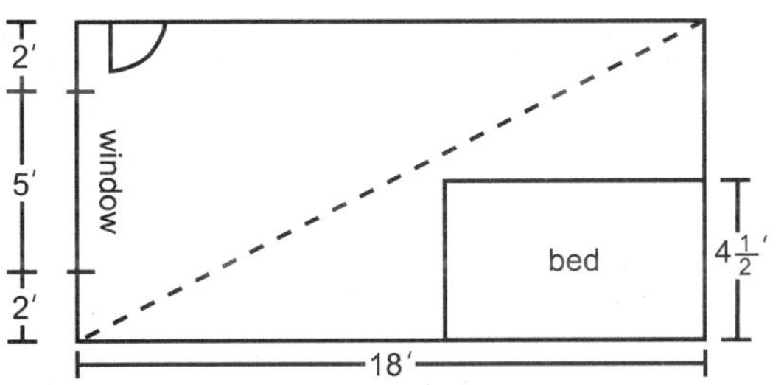

Which of the following questions **cannot** be answered using the given information?

A What is the area of Amanda's room?

B What is the perimeter of Amanda's room?

C What is the area of the window?

D What is the distance of the diagonal line across Amanda's room?

3. Justin played 13 games of football during the fall season and rushed a total of 145 yards. His team won 5 of the games in which Justin rushed at least 10 yards. Which of the following questions can be answered using the given information?

A What is the average number of yards Justin rushed each game?

B What is the average number of points the team scored during each game?

C What is the ratio of games won to games lost during the fall season?

D What is the total number of yards Justin rushed in the games his team won?

4. Cheyenne competes in a horse-riding competition. She finishes her race in a time that is 0.34 seconds slower than the third-place winner and 4.2 seconds slower than the first-place winner. The second-place winner finished the race in 3 minutes 12.2 seconds. Which of the following is necessary to determine Cheyenne's riding time?

A the difference in the first- and second-place times

B the place in which Cheyenne finished the race

C the difference in the first- and third-place times

D the amount of time deducted for each riding error

©ECS Learning Systems, Inc. All rights reserved

Notes

©ECS Learning Systems, Inc. All rights reserved

Mathematical Representation

IV. Express or solve problems using mathematical representation.

A. Represent multiplication and division situations involving fractions and decimals

B. Formulate a possible problem situation when given a simple equation

C. Solve problems by reading and interpreting collected data

©ECS Learning Systems, Inc. All rights reserved

Objective 4: Pretest

IV.A Represent multiplication and division situations involving fractions and decimals (1–6)

1. Mrs. Taylor earns \$732.98 each week. She pays her son's babysitter \$36 each week. Which equation can be used to determine what percent of her weekly pay Mrs. Taylor pays her babysitter?

A $\frac{732.98}{36} \times 100 =$

B $\frac{36}{732.98} \times 100 =$

C $100(732.98 - 36) =$

D $\frac{36 + 732.98}{100} =$

2. Keesha's puppy can eat an $8\frac{1}{2}$-pound bag of dog food in 18 days. She feeds the puppy twice a day. Which equation can be used to find f, the number of ounces of dog food Keesha's puppy eats each day?

A $f = (8\frac{1}{2} \times 16) \div 18$

B $f = \frac{8\frac{1}{2} \times 18}{16}$

C $f = 8\frac{1}{2} \div 16$

D $f = \frac{18}{8\frac{1}{2} \times 16}$

3. When full, a cereal box contains 26.9 ounces of cereal. The average serving of cereal is 4.5 ounces. Which equation can be used to find s, the total number of average servings that are in the box when it is full?

A $s = 4.5 \div 26.9$

B $s = (26.9 \div 16) \times 4.5$

C $s = 26.9 \times 4.5$

D $s = 26.9 \div 4.5$

4. Mark Twain Middle School's baseball game lasted for $2\frac{2}{3}$ hours. The teams played only 7 innings. Which equation can be used to determine the average amount of time it took the players to complete each inning?

A $2\frac{2}{3} \times 7 =$

B $2\frac{2}{3} \div 7 =$

C $7 \div 2\frac{2}{3} =$

D $\frac{2\frac{2}{3} \times 9}{7} =$

5. Virginia wrapped a present for her brother. The rectangular gift box was 27.3 centimeters long, 14.26 centimeters wide, and 5.5 centimeters high. Which equation can be used to find w, the total amount of wrapping paper Virginia needed to wrap the present?

A $w = (27.3 \times 14.26 \times 5.5)^2$

B $w = 2(27.3 \times 14.26 \times 5.5)$

C $w = 2(27.3 \times 14.26) + 2(14.26 \times 5.5) + 2(27.3 \times 5.5)$

D $w = 5.5(27.3 + 14.26) + 14.26(5.5 \times 27.3)$

©ECS Learning Systems, Inc.

All rights reserved

6. Dominic made wooden frames for 4 rectangular pictures that were each 12 inches long and $7\frac{1}{2}$ inches wide. Which equation can be used to find w, the total amount of wood Dominic needed to buy for framing all 4 pictures?

   A $w = 12 + 7\frac{1}{2}$

   B $w = 4(7\frac{1}{2} \times 12)$

   C $w = 4(7\frac{1}{2} + 7\frac{1}{2}) + 4(12 + 12)$

   D $w = 4(12) + 7\frac{1}{2}$

IV.B Formulate a possible problem situation when given a simple equation (7–11)

7. Which problem situation can be solved with the following equation?

   $$s = 12t$$

   A What is s, the number of eggs in t dozens?

   B What is s, the number of ounces in t pounds?

   C What is s, the number of feet in t inches?

   D What is s, the number of days in t years?

8. In a basketball game, each field goal (a) earns the team 2 points; three-point shots (b) earn the team 3 points; and free throws (c) earn the team 1 point. Which equation best represents how to find p, the total points scored in a game of basketball?

   A $p = \frac{2 + 3 + 1}{a + b + c}$

   B $p = x(2 + 3 + 1)$

   C $p = 2a + 3b - c$

   D $p = 2a + 3b + c$

9. Which problem situation could be solved with the following equation?

   $$x = 52d$$

   A What is x, the number of sides on d octagons?

   B What is x, the number of weeks in d years?

   C What is x, the number of ounces in d pounds?

   D What is x, the number of days in d years?

©ECS Learning Systems, Inc.

All rights reserved

10. Which problem situation could be solved with the following equation?

$$y = 346(a \times b)$$

A What is y, the total circumference of 346 circles with a radius of a and a diameter of b?

B What is y, the total perimeter of 346 rectangles with a height of a and a width of b?

C What is y, the total area of 346 rectangles with a height of a and a width of b?

D What is y, the total area of 346 trapezoids with base lengths of a and b?

11. Which problem situation could be solved with the following equation?

$$t = 3.45s$$

A What is t, the perimeter of a rectangle with a height of 3.45 inches and a width of s?

B What is t, the area of a circle with a diameter of 3.45 and a radius of s?

C What is t, the area of a rectangle with a height of 3.45 and a width of s?

D What is t, the circumference of a circle with a radius of 3.45 and a diameter of s?

IV.C Solve problems by reading and interpreting collected data (12–17)

The graph below shows the number of centimeters 2 different plants grew over a period of 6 months. Use the graph to answer questions 12–13.

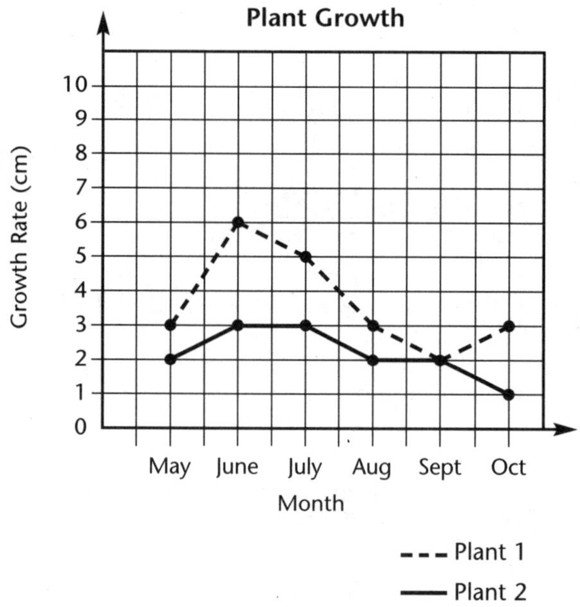

12. On average, how many centimeters per day did Plant 1 grow in June?

A 5 cm

B 2 cm

C 0.5 cm

D 0.2 cm

13. During which month did the 2 plants grow at the same rate?

A May

B July

C August

D September

©ECS Learning Systems, Inc.

All rights reserved

George Washington Middle School built a new gymnasium for their basketball team. The original budget for the gymnasium was $1 million. The pie graph below shows how the budget was distributed. Use the information on the pie graph to answer questions 14–17.

Expenditures for New Gymnasium

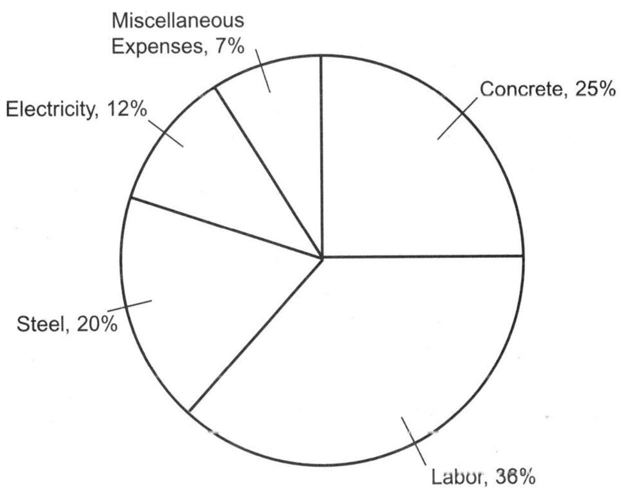

14. How much did the school spend on concrete, steel, and labor?

A \$810,000

B \$560,000

C \$450,000

D \$80,000

15. How much more money was spent on concrete than steel?

A \$5,000,000

B \$50,000

C \$20,000

D \$5,000

16. How much money did the school spend on electricity and miscellaneous expenses?

A \$19,000

B \$50,000

C \$190,000

D \$200,000

17. If George Washington Middle School ended up spending a total of \$1,250,000 on the gymnasium, how much money would have been spent on concrete?

A \$31,250

B \$50,000

C \$250,000

D \$312,500

©ECS Learning Systems, Inc.

All rights reserved

Practice 4.A1

IV.A Represent multiplication and division situations involving fractions and decimals

1. Mrs. Meeks is a professional musician. She charges $145.50 to play the piano at formal events and an additional $40.25 to play during any rehearsals. In 1 month, Mrs. Meeks played the piano at 3 formal events and 2 of the rehearsals. Which equation can be used to determine T, the total amount of money that she made that month?

   A $2(145.5) + 3(40.25) = T$

   B $3(145.5) + 2(40.25) = T$

   C $5(145.5 + 40.25) = T$

   D $\frac{145.5 + 40.25}{2 + 3} = T$

2. Marla is $48\frac{3}{4}$ years old. She has spent an equal number of years living in Sante Fe, Des Moines, and Raleigh. Which equation can be used to determine Y, the number of years Marla lived in each city?

   A $48\frac{3}{4} \div 12 = Y$

   B $48\frac{3}{4} \div 6 = Y$

   C $48\frac{3}{4} \div 3 = Y$

   D $48\frac{3}{4} \times 3 = Y$

3. Hosea went to the movies 4 times in May. Three of the times, a ticket cost $5.50. The last time, Hosea paid the matinee price for a ticket, which was $4.25. Which equation can be used to find m, the total amount of money Hosea spent on movie tickets in May?

   A $m = 3(5.5 + 4.25)$

   B $m = \frac{5.5}{3} + 4.25$

   C $m = 3(5.5) + 4.25$

   D $4.25 = (5.5)m - 3$

4. There are 4,325 students enrolled in Durham High School, 8% of which are in the school's marching band. Half of the band members are seniors. Which equation can be used to find s, the number of band members who are seniors?

   A $\frac{4,325 \times 0.08}{2} = s$

   B $\frac{4,325 \times 8}{2} = s$

   C $s = \frac{0.08}{2} + 4,325$

   D $s = \frac{4,325}{2} \div 0.08$

©ECS Learning Systems, Inc.

All rights reserved

Practice 4.A2

IV.A Represent multiplication and division situations involving fractions and decimals

1. Penelope shopped at the market 3 times this week. On Wednesday, she bought $2\frac{1}{2}$ pounds of apples for 67¢ a pound and $\frac{3}{4}$ pound of cucumbers for 99¢ a pound. Which equation can be used to determine C, the total cost of Penelope's purchases on Wednesday?

   A $\frac{(2\frac{1}{2} \times 0.67) + (\frac{3}{4} \times 0.99)}{3} = C$

   B $0.67 + 0.99 = C$

   C $(2\frac{1}{2} \times 0.67) + (\frac{3}{4} \times 0.99) = C$

   D $(2\frac{1}{2} \times \frac{3}{4}) + (0.67 + 0.99) = C$

2. Darius spends $1\frac{1}{2}$ hours in gymnastics practice each week. He also spends $2\frac{2}{3}$ hours in baseball practice each week. Which equation can be used to determine the amount of time Darius spends in gymnastics and baseball practice each month (4 weeks)?

   A $\frac{(1\frac{1}{2} + 2\frac{2}{3})}{4} =$

   B $4(1\frac{1}{2} + 2\frac{2}{3}) =$

   C $4 \times 2\frac{2}{3} =$

   D $1\frac{1}{2} + 2\frac{2}{3} =$

3. Mr. Gennisy is the contractor responsible for building a 37-story building. He estimates that the cost of wiring electrical lines to the building will be $12,356.15. Which equation can be used to find x, the average cost of wiring each floor?

   A $\frac{37}{12,356.15} = x$

   B $37 = \frac{x}{12,356.15}$

   C $\frac{x}{37} = 12,356.15$

   D $x = 12,356.15 \div 37$

4. Joshua can swim 30-meter laps at a rate of 4 laps per minute. He usually swims for $1\frac{1}{2}$ hours each day. Which equation can be used to find M, the number of meters Joshua swims per second?

   A $\frac{30 \times 4}{60} = M$

   B $30 \div 4 = M$

   C $\frac{30 \times 4}{1\frac{1}{2}} = M$

   D $30 \times 4 = M \div 60$

©ECS Learning Systems, Inc.

All rights reserved

Practice 4.A3

IV.A Represent multiplication and division situations involving fractions and decimals

1. Wendy took 3 dimes, 2 nickels, and 7 pennies out of her pocket. Which equation can be used to determine M, the amount of money Wendy has?

   A $3(10) + 2(0.05) + 7(0.01) = M$

   B $(3 + 2 + 7) \times (10 + 5 + 1) = M$

   C $3(0.1) + 2(0.05) + 7(0.01) = M$

   D $7(0.1) + 2(0.05) \times 3(0.01) = M$

2. Mika is an event planner. For a fundraiser dinner, she spent a total of \$458.50 for 78 candles and 32 silk flowers. The total cost of the flowers was \$50.56. Which equation can be used to determine C, the cost of each candle?

   A $458.5 - (50.56 \div 32) = C$

   B $C = 458.5 \div (78 \times 50.56)$

   C $C = \frac{458.5 - 50.56}{78}$

   D $C = 458.5 - (78 - 32)$

3. Greg's apartment is 839.8 square feet. The area of his storage facility is $\frac{2}{3}$ the area of his apartment. Which equation can be used to find A, the area of Greg's storage facility?

   A $839.8 \div 2\frac{2}{3} = A$

   B $839.8 \times \frac{2}{3} = A$

   C $A = 839.8 \times \frac{1}{3}$

   D $839.8 = \frac{2}{3}A$

4. Starry Oaks subdivision is in the process of being built. Of the 34 blocks in the subdivision, $\frac{1}{2}$ are designed to have 12 houses on them. The remaining blocks are designed to have 10 houses on them. Which equation can be used to determine the total number of houses that can be built in Starry Oaks subdivision?

   A $\frac{(34 \times 12)}{2} + \frac{(34 \times 10)}{2} =$

   B $\frac{12 + 10}{34} =$

   C $34(12 + 10) =$

   D $\frac{2 \times 34}{12 + 10} =$

©ECS Learning Systems, Inc.

All rights reserved

Practice 4.A4

IV.A Represent multiplication and division situations involving fractions and decimals

1. Mr. Dutcher has \$4,679.88 left to pay on his car loan. He plans to pay off the car loan in 12 months. Which equation can be used to determine *P*, the average amount Mr. Dutcher must pay on his car loan each month in order to pay it off in 12 months?

A $4{,}679.88 \times 12 = P$

B $\frac{12}{4{,}679.88} = P$

C $4{,}679.88 \div 12 = P$

D $4{,}679.88P = 12$

2. Allison is a chef at an Italian restaurant. One night, she used the following amounts of parmesan cheese in 3 different dishes: $2\frac{1}{2}$ tablespoons, $4\frac{5}{6}$ tablespoons, and $\frac{2}{3}$ tablespoons. Which equation can be used to find *C*, the average amount of parmesan cheese Allison used in each dish?

A $2\frac{1}{2} + 4\frac{5}{6} + \frac{2}{3} = C$

B $\frac{2\frac{1}{2} + 4\frac{5}{6} + \frac{2}{3}}{3} = C$

C $4(2\frac{1}{2} + 4\frac{5}{6} + \frac{2}{3}) = C$

D $\frac{3}{2\frac{1}{2} + 4\frac{5}{6} + \frac{2}{3}} = C$

3. Demitri receives \$4.25 in allowance each week. He must save \$2, but is allowed to spend the remaining money as he chooses. Which equation can be used to find *m*, the amount of money Demitri is allowed to spend over a 3-week period?

A $(4.25 - 2) \times 3 = m$

B $\frac{4.25}{3} = m$

C $4.25(2m) = 3$

D $4.25 - \frac{2}{3} = m$

4. Jessica's horse, Jumper, can jump over a fence that is $\frac{1}{2}$ his height. Which equation can be used to determine *H*, the height Jumper can jump if he is $9\frac{2}{3}$ feet tall?

A $9\frac{2}{3} \div \frac{1}{2} = H$

B $\frac{2}{9\frac{2}{3}} = H$

C $9\frac{2}{3} \times 2 = H$

D $9\frac{2}{3} \div 2 = H$

©ECS Learning Systems, Inc.

All rights reserved

Practice 4.A5

IV.A Represent multiplication and division situations involving fractions and decimals

1. Derrick's dog, Blackie, has a litter of 9 puppies. There are 3 females in the litter. Also, $\frac{2}{3}$ of the puppies are black and $\frac{1}{3}$ are brown with black and white spots. Which equation can be used to find p, the number of black puppies?

   A $p = 9 \times \frac{1}{3}$

   B $p = (9 - 3) \times \frac{2}{3}$

   C $p = 9 \times \frac{2}{3}$

   D $p = 9 \div \frac{2}{3}$

2. Victor spends a total of 62 hours working at 2 different jobs each week. He spends 65% of that time working at his first job. Which equation can be used to find h, the number of hours he spends working at his second job each week?

   A $\frac{62}{0.65} = h$

   B $62 - (62 \times 0.65) = h$

   C $65 - 62 = h$

   D $h = 62 \times 0.65$

3. Marianna is a server at a restaurant. She earns an average of 16% of her total sales in tips each night. If Marianna sells $459.19 on Friday night, which equation can be used to determine T, the amount of money she earns in tips?

   A $459.19 \div 16 = T$

   B $459.19 + (459.19 \times 0.16) = T$

   C $459.19 \times 16 = T$

   D $459.19 \times 0.16 = T$

4. Valerie needs a total of $4\frac{1}{2}$ cups of sugar to make 3 batches of cookies. Each batch produces 42 cookies. Which equation can be used to determine S, the average amount of sugar in each cookie?

   A $\frac{4\frac{1}{2}}{3 \times 42} = S$

   B $\frac{3 \times 42}{4\frac{1}{2}} = S$

   C $\frac{42}{3} \times 4\frac{1}{2} = S$

   D $\frac{42}{4\frac{1}{2}} = S$

©ECS Learning Systems, Inc.

All rights reserved

Practice 4.A6

IV.A Represent multiplication and division situations involving fractions and decimals

1. Isaiah draws cartoon pictures for fun. He can draw $2\frac{1}{2}$ cartoons in 1 hour. It takes him an additional 10 minutes per cartoon to write in the words. Which equation shows how to determine *M*, the number of minutes it takes Isaiah to draw and write words for 1 cartoon?

A $\dfrac{1 + 10}{2\frac{1}{2}} = M$

B $\dfrac{60}{2\frac{1}{2}} + 10 = M$

C $2\frac{1}{2} \times (60 + 10) = M$

D $\dfrac{2\frac{1}{2} + 10}{60} = M$

2. Mrs. Garcia uses her new printer to print $\frac{2}{3}$ of a book that is 1,288 pages long. Her printer can print 4,500 pages before it needs a new cartridge. Which equation can be used to find *x*, how many more pages Mrs. Garcia can print before the printer needs a new cartridge?

A $\frac{2}{3} = 4{,}500(1{,}288) \div x$

B $x = 4{,}500(\frac{2}{3} \times 1{,}288)$

C $x = 4{,}500 - (1{,}288 \times \frac{2}{3})$

D $4{,}500 = 1{,}288 - \frac{2}{3}x$

3. Jocelyn rides the bus to school $\frac{3}{4}$ of the school year. The rest of the year, she rides her bicycle. If there are 182 days in the school year, which equation can be used to find *d*, the number of days Jocelyn rides her bicycle to school?

A $\frac{3}{4}(182) = d$

B $\frac{1}{4}(182) = d$

C $d = \dfrac{182 \times 4}{3}$

D $d = 182 \div \frac{1}{4}$

4. Fernando works part time at a pet shop. If he sells more than \$400 in the first week, he will break the store sales record for part-time workers. In the first 2 weeks, Fernando sells 12 pets for a total of \$265.34, dog food worth \$32.50, and birdseed worth \$13.12. Which equation can be used to find *M*, the average amount of merchandise Fernando sold each week?

A $400 - (265.34 + 32.5 + 13.12) - M$

B $\dfrac{265.34}{12} + 32.5 + 13.12 = M$

C $2(265.34 + 32.5 + 13.12) = M$

D $\dfrac{265.34 + 32.5 + 13.12}{2} = M$

©ECS Learning Systems, Inc.

All rights reserved

Practice 4.B1

IV.B Formulate a possible problem situation when given a simple equation

1. Which problem situation could be solved with the following equation?

$$y = 24x$$

A What is y, the number of feet in x miles?

B What is y, the number of cups in x gallons?

C What is y, the number of hours in x days?

D What is y, the number of seconds in x weeks?

2. Which problem situation could be solved with the following equation?

$$d = 22(e \times 7)$$

A What is d, the area of 22 obtuse triangles with a height of 7 inches?

B What is d, the area of a trapezoid that is 7 inches high and 22 inches wide?

C What is d, the perimeter of 22 octagons with sides that are 7 inches long?

D What is d, the total area of 22 rectangles with a length of e inches and a width of 7 inches?

3. Mr. Jackson's bakery charges $4 per cake and 25¢ per cookie. Which equation best represents how to find y, the amount of money Mr. Jackson's bakery makes from selling cakes, x, and cookies, z?

A $y = 4x + 0.25z$

B $y = (4 \times 0.25)xz$

C $y = 4(xz)(0.25)$

D $y = 4x + 25z$

4. Which problem situation could be solved with the following equation?

$$q = 100r$$

A What is q, the number of fluid ounces in r gallons?

B What is q, the number of ounces in r tons?

C What is q, the number of centimeters in r meters?

D What is q, the number of milliseconds in r hours?

©ECS Learning Systems, Inc.

All rights reserved

Practice 4.B2

IV.B Formulate a possible problem situation when given a simple equation

1. Which problem situation could be solved with the following equation?

$$y = \frac{x}{365}$$

A What is y, the number of inches in 365 yards?

B What is y, the number of days in x years?

C What is y, the number of pounds in 365 tons?

D What is y, the number of years in x days?

2. Which problem situation could be solved with the following equation?

$$m = 50 + 7x$$

A Peter earns m dollars for each yard he mows and \$7 for each hedge he trims. What is x, the total dollar amount he earns?

B Peter earns \$50 each month for mowing yards and pays \$7 each time he gets his car washed. What is m, the total dollar amount he earns?

C Peter earns \$50 a day plus \$7 for working x overtime hours. What is m, the total dollar amount he earns?

D Peter mows 50 yards each month, and he earns \$7 per yard. What is m, the total dollar amount he earns?

3. Which problem situation could be solved with the following equation?

$$y = 22(2 \times 3.14 \times 4)$$

A What is y, the total circumference of 22 circles with a radius of 4?

B What is y, the total circumference of 22 circles with a diameter of 4?

C What is y, the total perimeter of 22 squares with a radius of 4?

D What is y, the total area of 22 circles with a radius of 4?

4. Which problem situation could be solved with the following equation?

$$y = 6 \times 7$$

A What is y, the area of Jennifer's rectangular room that has a 6-foot diameter and a 7-foot circumference?

B What is y, the area of Jennifer's rectangular room that has an obtuse angle measuring 6° and an acute angle measuring 7°?

C What is y, the area of Jennifer's rectangular room that is 6 feet wide and 7 feet long?

D What is y, the perimeter of Jennifer's rectangular room that is 6 feet wide and 7 feet long?

©ECS Learning Systems, Inc.

All rights reserved

Practice 4.B3

IV.B Formulate a possible problem situation when given a simple equation

1. Which problem situation could be solved with the following equation?

$$y = 2x + 2z$$

A What is y, the area of 2 circles with a diameter of x and a radius of z?

B What is y, the perimeter of a rectangle with sides x and z?

C What is y, the area of a triangle with a base length of x and a height of z?

D What is y, the area of a rectangle with sides x and z?

2. Which problem situation could be solved with the following equation?

$$y = \frac{x}{36}$$

A What is y, the number of yards in x inches?

B What is y, the number of centimeters in x meters?

C What is y, the number of ounces in x cups?

D What is y, the number of inches in x yards?

3. A total of 7 circles each have a diameter of x. Which equation best represents how to find c, the total circumference of all 7 circles?

A $c = 7(7x)$

B $c = 3.14(7) + x$

C $c = 3.14x + 7$

D $c = 7(3.14x)$

4. Which problem situation could be solved with the following equation?

$$25 + 2x = y$$

A Tony earns a flat rate of \$25 per day parking cars. He also earns \$2 per car he parks. What is y, the total amount Tony earns parking cars per day?

B Tony earns a flat rate of \$2 per day parking cars. He also earns \$25 per car he parks. What is y, the total amount Tony earns parking cars per day?

C Tony earns a flat rate of x dollars per day parking cars. He also earns \$25 per car he parks. What is y, the total amount Tony earns parking cars per day?

D Tony earns a flat rate of \$25 per day parking cars. He also pays \$2 in taxes for each car he parks. What is y, the total amount Tony earns parking cars per day?

©ECS Learning Systems, Inc.

All rights reserved

Practice 4.B4

IV.B Formulate a possible problem situation when given a simple equation

1. Which equation could help Sally determine S, the number of seconds in x days?

A $S = x + (60 \times 60)$

B $S = x(60 \times 60)$

C $S = (60 \times 60) + (x + 24)$

D $S = x(60 \times 60 \times 24)$

2. Which problem situation could be solved with the following equation?

$$g = 4x$$

A What is g, the area of a rectangle with a height of 4 inches?

B What is g, the area of a square with each side equal to x?

C What is g, the area of a circle with a radius equal to x?

D What is g, the perimeter of a square with each side equal to x?

3. Which problem situation could be solved with the following equation?

$$d = \frac{e}{8}$$

A What is d, the number of cups in e fluid ounces?

B What is d, the number of ounces in e pounds?

C What is d, the number of inches in e feet?

D What is d, the number of gallons in e liters?

4. Stefan wants to make necklaces for his friends. Each girl's necklace, y, needs 74 beads, and each boy's necklace, x, needs 95 beads. Which equation best represents how to find B, the number of beads Stefan must purchase to make necklaces for 1 boy and 1 girl?

A $B = 95x + 74y$

B $B = \frac{95 + 74}{x + y}$

C $B = x(95 + 74)$

D $B = (95 - 74) \times (x + y)$

©ECS Learning Systems, Inc. All rights reserved

Practice 4.B5

IV.B Formulate a possible problem situation when given a simple equation

1. Which problem situation could be solved with the following equation?

$$y = 3.14 \times 8$$

A What is y, the circumference of 8 circles?

B What is y, the circumference of 1 circle with a radius of 8 inches?

C What is y, the circumference of 1 circle with a diameter of 8 inches?

D What is y, the area of an octagon with each side measuring 8 inches?

2. Nick and Eric combine and save their allowances each week to purchase a new video game. Eric receives $15 per week and Nick receives $10 per week. Which equation best represents how to find x, the number of weeks the boys must save in order to purchase a video game that costs $75?

A $x = 75 - (15 + 10)$

B $75 = x(15 \times 10)$

C $x = 75(15 + 10)$

D $x = \frac{75}{15 + 10}$

3. Which problem situation could be solved with the following equation?

$$10d + 100 = m$$

A Jill earns $10 each day she works for her uncle plus a flat rate of $100 a month. What is m, the total dollar amount Jill earns in 1 month when she works d number of days?

B Jill earns $100 each day she works for her uncle plus a flat rate of $10 a month. What is m, the total dollar amount Jill earns in 1 month?

C Jill earns $10 for every purchase she sells at her uncle's store. What is m, the total dollar amount Jill earns in 100 months?

D Jill earns d amount of dollars for every 10 items she sells. She also makes $100 per day working at her uncle's store. What is m, the total dollar amount Jill earns in 1 year?

4. Which problem situation could be solved with the following equation?

$$s = 100t$$

A What is s, the number of days in t centuries?

B What is s, the number of years in t decades?

C What is s, the number of years in t millenniums?

D What is s, the number of years in t centuries?

©ECS Learning Systems, Inc. All rights reserved

Practice 4.B6

IV.B Formulate a possible problem situation when given a simple equation

1. Which problem situation could be solved with the following equation?

$$y = 8(6 \times b)$$

A What is y, the total area of 8 rectangles with a height of 6 inches and a width of b?

B What is y, the perimeter of an octagon whose sides equal 6 inches?

C What is y, the total area of 8 triangles with a height of 6 inches and a width of b?

D What is y, the area of a rectangle with a height of 8 inches and a width of 6 inches?

2. Dylan earned $10 allowance each week. He also earned $25 for each of the 8 yards he mowed in the last 4 weeks. Which equation best represents how to find m, the total amount of money Dylan made in the last 4 weeks?

A $m = 10 + 24 \times 4$

B $m = 10(25 \times 8) + 4$

C $m = 4(10) + 25$

D $m = (4 \times 10) + (8 \times 25)$

3. Which problem situation could be solved with the following equation?

$$y = \frac{3x}{2}$$

A What is y, the total area of 3 triangles with a base of x meters and a height of 2 meters?

B What is y, the perimeter of a square with 4 sides each equal to 3 meters?

C What is y, the total area of a triangle with a height of 3 meters and a base equal to x?

D What is y, the perimeter of a triangle with a height of 3 meters and a base equal to x?

4. Mrs. Murphy's car can travel 287 miles on 1 tank of gasoline. The gas tank holds 12 gallons of gasoline. Which equation best represents how to find m, the number of miles Mrs. Murphy's car can travel per gallon of gasoline?

A $m = 287 \times 12$

B $m = 287 \div 12$

C $\frac{287}{1} = \frac{m}{12}$

D $m - \frac{12}{287}$

©ECS Learning Systems, Inc. All rights reserved

Practice 4.B7

IV.B Formulate a possible problem situation when given a simple equation

1. Twice each day, Danielle runs around a circular track with a radius of 25 meters. Which equation best represents how to find m, the number of meters Danielle runs in x days?

   A $m = x(25 \times 2)$

   B $m = 25(3.14x)$

   C $m = x(3.14 \times 25)$

   D $m = x(3.14 \times 50)$

2. Which problem situation could be solved with the following equation?

   $$y = 3(4x)$$

   A What is y, the circumference of 3 squares with each side equal to x?

   B What is y, the circumference of a circle with a radius of $4x$?

   C What is y, the area of a square with each side equal to 4?

   D What is y, the area of a rectangle with a height of x inches and a width of 3 inches?

3. Which problem situation could be solved with the following equation?

   $$y = 2{,}000 \times 16 \times b$$

   A What is y, the number of seconds in b hours?

   B What is y, the number of ounces in b tons?

   C What is y, the number of milli-meters in b meters?

   D What is y, the number of feet in b yards?

4. Which problem situation could be solved with the following equation?

   $$t = 376(3.14u^2)$$

   A What is t, the total area of 376 squares with the length of each side equal to u?

   B What is t, the total area of 376 circles with a radius of u?

   C What is t, the total perimeter of 376 trapezoids with a height of u?

   D What is t, the total circumference of 376 circles with a radius of u?

©ECS Learning Systems, Inc. All rights reserved

Practice 4.B8

IV.B Formulate a possible problem situation when given a simple equation

1. Kari paints x half-circles, each with a radius of 9 centimeters. She needs to know if she has enough paint to color in the circles. Which equation best represents how to find A, the area of the half-circles? ($\pi = 3.1$)

   A $x = A(3.1 \times 9)$

   B $A = \frac{3.1x^2}{2}$

   C $A = x\left(\frac{3.1 \times 9^2}{2}\right)$

   D $A = x(3.1 \times 9)^2$

2. Which problem situation could be solved with the following equation?

   $$y = 2(4 \times m)$$

   A What is y, the perimeter of a rectangle with a height of 4 meters and a width of m?

   B What is y, the total area of 2 rectangles with a height of 4 meters and a width of m?

   C What is y, the total area of 2 rectangles with a height of 8 meters and a width of $2m$?

   D What is y, the volume of a cylinder with a height of 2 meters and a radius of m?

3. Which problem situation could be solved with the following equation?

   $$y = x + (12 \times 6)$$

   A What is y, the surface area of Jill's original rectangular swimming pool, x, plus the area of a new rectangular pool that is 12 yards long and 6 yards wide?

   B What is y, the area of Jill's original rectangular swimming pool with a length of 12 yards and a width of 6 yards?

   C What is y, the area of all of Jill's swimming pools with heights of x and widths of 12 or 6 yards?

   D What is y, the perimeter of Jill's rectangular swimming pool, x, plus the perimeter of a new rectangular pool that is 12 yards long and 6 yards wide?

4. Which problem situation could be solved with the following equation?

   $$y = 3.14 \times 3^2 \times k$$

   A What is y, the area of a circle with a radius of 3 centimeters?

   B What is y, the surface area of a cube with each side equal to k?

   C What is y, the volume of a cylinder with a radius of 3 centimeters and a height of k?

   D What is y, the surface area of a cylinder with a radius of 3 centimeters and a height of k?

©ECS Learning Systems, Inc. All rights reserved

Practice 4.C1

The graph below shows how many orders were placed for different menu items at John Adams High School cafeteria in April 2001. Use the graph to answer questions 1–4.

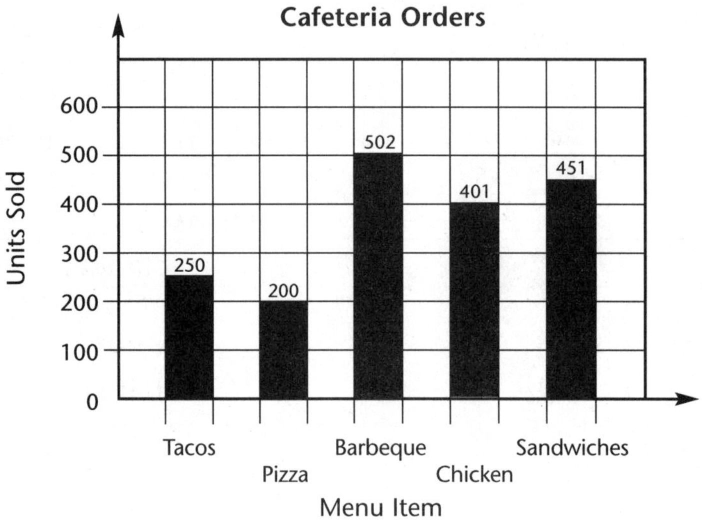

IV.C Solve problems by reading and interpreting collected data

1. The cafeteria manager decides to remove pizza from the menu. He is sure it will increase the number of tacos sold by 30%. If the cafeteria manager's prediction is correct, how many tacos will be sold in May 2001?

   A 325

   B 315

   C 175

   D 75

2. What percent of the overall cafeteria sales comes from sandwiches?

   A 4%

   B 25%

   C 30%

   D 40%

3. According to the chart, which menu item has a sales rate that is almost exactly 100% greater than the number of tacos sold?

   A Pizza

   B Barbeque

   C Chicken

   D Sandwiches

4. Ninth-graders purchased 40% of the tacos sold in April. How many tacos did the ninth-graders purchase?

   A 1,000

   B 625

   C 100

   D 80

©ECS Learning Systems, Inc. All rights reserved

Practice 4.C2

The graph below shows the source of Paul Robinson's $800 savings over the past year. Use the graph to answer questions 1–4.

Paul Robinson's Yearly Savings

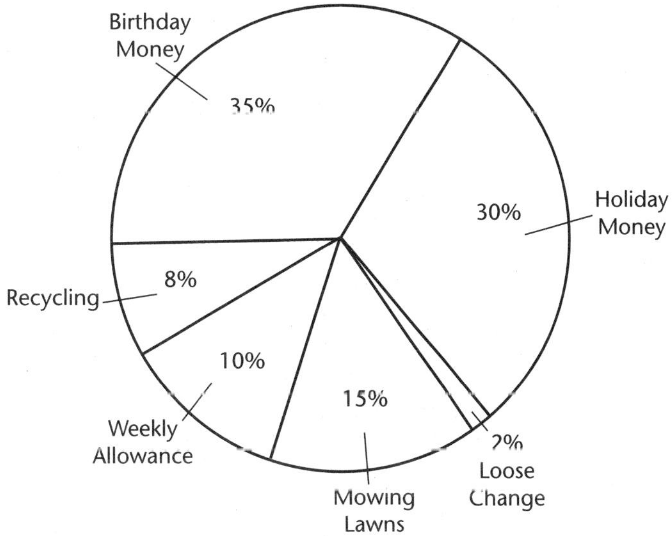

IV.C Solve problems by reading and interpreting collected data

1. How much of Paul's savings came from holiday and birthday money combined?

   A $240

   B $256

   C $480

   D $520

2. What percentage of Paul's savings came from sources other than holiday and birthday money?

   A 33%

   B 35%

   C 65%

   D $280

3. If Paul had saved only $500 last year and his source percentages were exactly the same, how much of his savings would have come from loose change and his weekly allowance?

   A $40

   B $60

   C $96

   D $440

4. If Paul had not saved any money from the loose change he found, how much money would be in his savings account?

   A $784

   B $640

   C $160

   D $16

©ECS Learning Systems, Inc. All rights reserved

Practice 4.C3

The graph below shows the number of computers sold by Ace Computer Company from the years 1996–2000. Use the graph to answer questions 1–4.

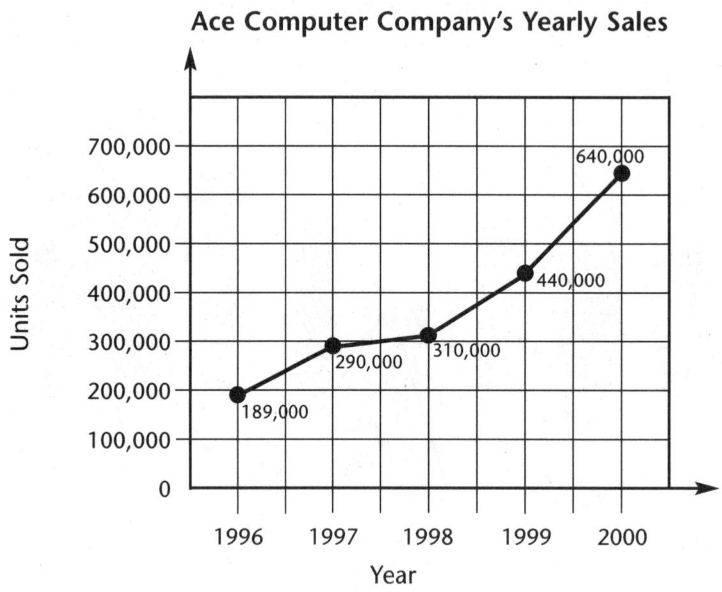

IV.C Solve problems by reading and interpreting collected data

1. Which year showed the greatest increase in the number of computers sold by Ace Computer Company?

   A 2000

   B 1999

   C 1998

   D 1996

2. Between which 2 years did the number of computers sold change the least?

   A 1996–1997

   B 1997–1998

   C 1998–1999

   D 1999–2000

3. What was the average number of computers sold each year over the 5-year period?

   A 1,869,000

   B 374,000

   C 373,800

   D 37,380

4. What was the median number of computers sold over the 5-year period?

   A 290,000

   B 300,000

   C 310,000

   D 373,800

©ECS Learning Systems, Inc. All rights reserved

Practice 4.C4

The graph below shows what percent of the student body approved of Principal Smith over a period of 8 years. Use the graph to answer questions 1–4.

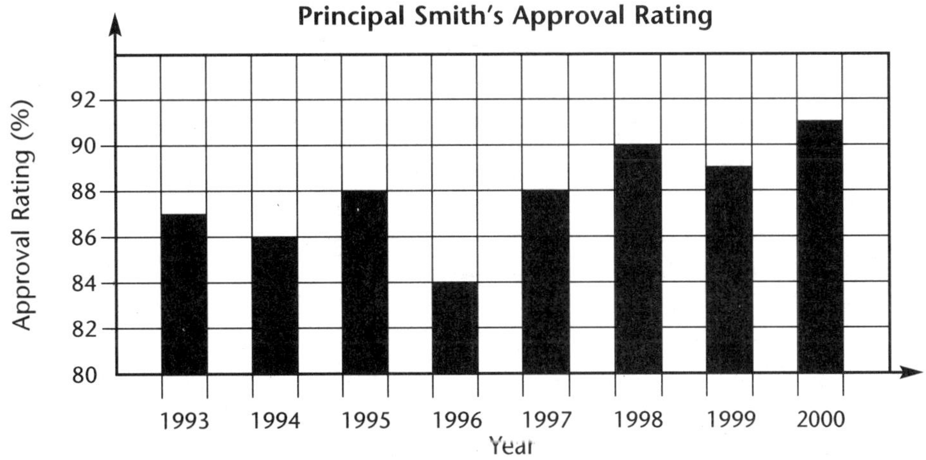

IV.C Solve problems by reading and interpreting collected data

1. If Principal Smith had 1,425 students at his school in 1997, how many approved of him?

   A 1,254

   B 1,154

   C 454

   D 88

2. What was the greatest percentage increase in Principal Smith's approval rating from one year to the next?

   A 1994–1995

   B 1995–1996

   C 1996–1997

   D 1999–2000

3. In both 1999 and 2000, 1,500 students attended Principal Smith's school. How many more students approved of him in the year 2000 than in 1999?

   A 20

   B 25

   C 30

   D 1,365

4. Next year, Principal Smith will move to a new school of 900 students. If his approval rating remains the same as it was in the year 2000, how many students will approve of him at the new school?

   A 91

   B 819

   C 820

   D 1,820

©ECS Learning Systems, Inc.

All rights reserved

Practice 4.C5

The graph below shows what percent of sales came from different baked goods at Brother's Bakery in June 2002. Use the graph to answer questions 1–4.

Brother's Bakery
June 2002 Sales Breakdown

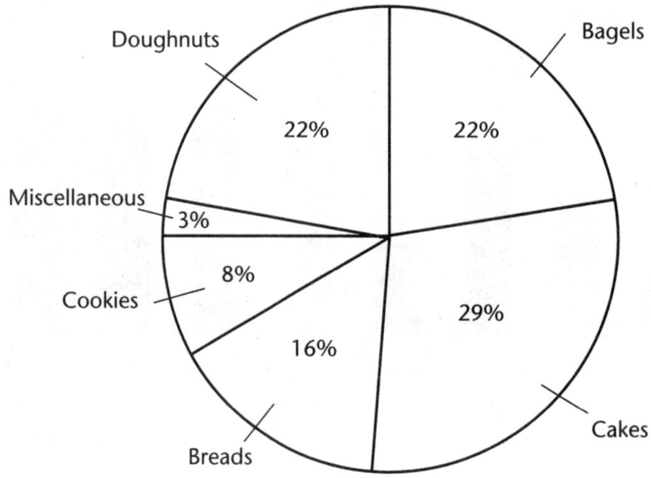

IV.C Solve problems by reading and interpreting collected data

1. If the bakery sold $23,769 in June, how much money did it make from bread and bagel sales?
   - A $5,229.18
   - B $6,655.32
   - C $8,032.22
   - D $9,032.22

2. During the holiday season, the owner expects to sell 50% more cookies than he sold in June 2002. If the bakery sold 2,000 cookies in June, how many cookies does the owner expect to sell during the holiday season?
   - A 3,000
   - B 2,100
   - C 2,050
   - D 1,000

3. In July 2002, bread sales at the bakery increased by 50%. Cake sales decreased. If the percentage of sales in other categories remained the same, what percentage of the bakery's total sales came from cake sales in July 2002?
   - A 11%
   - B 21%
   - C 24%
   - D 37%

4. At Brother's Bakery, how much of the $23,769 came from doughnut, bread, and cookie sales in June 2002?
   - A $9,032.22
   - B $10,933.74
   - C $11,533.74
   - D $12,835.26

©ECS Learning Systems, Inc. All rights reserved

Practice 4.C6

The graph below shows how many hip hop and rock CDs the Bandwidth Music Store sold in the year 2002. Use the graph to answer questions 1–4.

Bandwidth Music Store Sales, Hip Hop vs. Rock, 2002

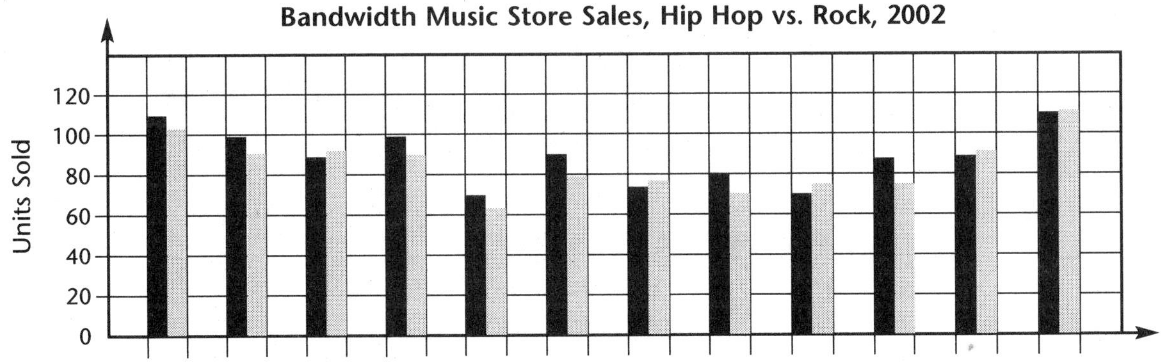

	Jan	Feb	Mar	Apr	May	Jun	Jul	Aug	Sep	Oct	Nov	Dec
■ Hip Hop	108	99	88	99	68	89	74	80	69	87	88	110
□ Rock	103	91	93	89	63	79	77	69	75	75	92	112

Months

IV.C Solve problems by reading and interpreting collected data

1. What was the average number of hip hop CDs the Bandwidth Music Store sold each month?

   A 88.25

   B 88.3

   C 89.92

   D 90

2. In which month was there the greatest difference in the number of hip hop and rock CDs sold at Bandwidth Music Store?

   A January

   B April

   C August

   D October

3. In which month was there the least difference in the number of hip hop and rock CDs sold at Bandwidth Music Store?

   A May

   B July

   C November

   D December

4. The owner of the store expects to sell 10% more rock CDs in January 2003 than he did in December 2002. Rounded to the nearest whole number, how many rock CDs does the owner expect to sell in January 2003?

   A 111

   B 113

   C 123

   D 224

©ECS Learning Systems, Inc.

All rights reserved

Practice 4.C7

The graph below shows the percentage of households with Internet access and personal computers over a period of 5 years. Use the graph to answer questions 1–4.

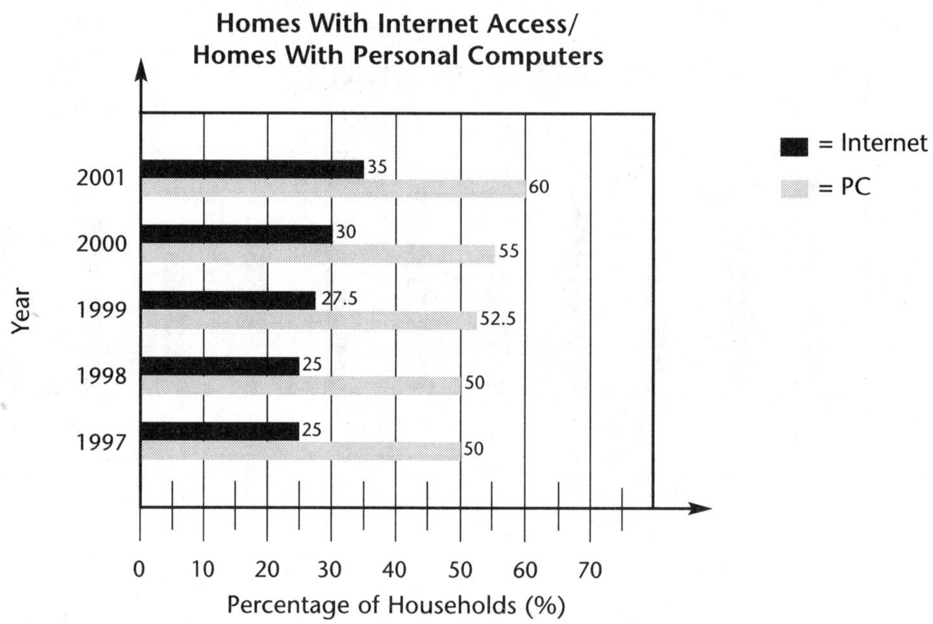

IV.C Solve problems by reading and interpreting collected data

1. If 567,000 households were polled in the year 1998, how many of them had Internet access?

   A 25

   B 131,750

   C 141,750

   D 425,250

2. According to the graph, the greatest increase in the number of households with Internet access took place between the years—

   A 1997 and 1998

   B 1998 and 1999

   C 1999 and 2000

   D 2000 and 2001

3. According to the graph, the number of households with personal computers increased the least between the years—

   A 1997 and 1998

   B 1998 and 1999

   C 1999 and 2000

   D 2000 and 2001

4. In the year 2001, there were 967,000 households polled. How many of these households had personal computers?

   A 580,200

   B 540,200

   C 531,850

   D 338,450

©ECS Learning Systems, Inc. All rights reserved

Practice 4.C8

The graph below shows the percentage of households that owned DVD players and VCRs starting in 1997 and predicting future sales trends through 2007. Use the graph to answer questions 1–4.

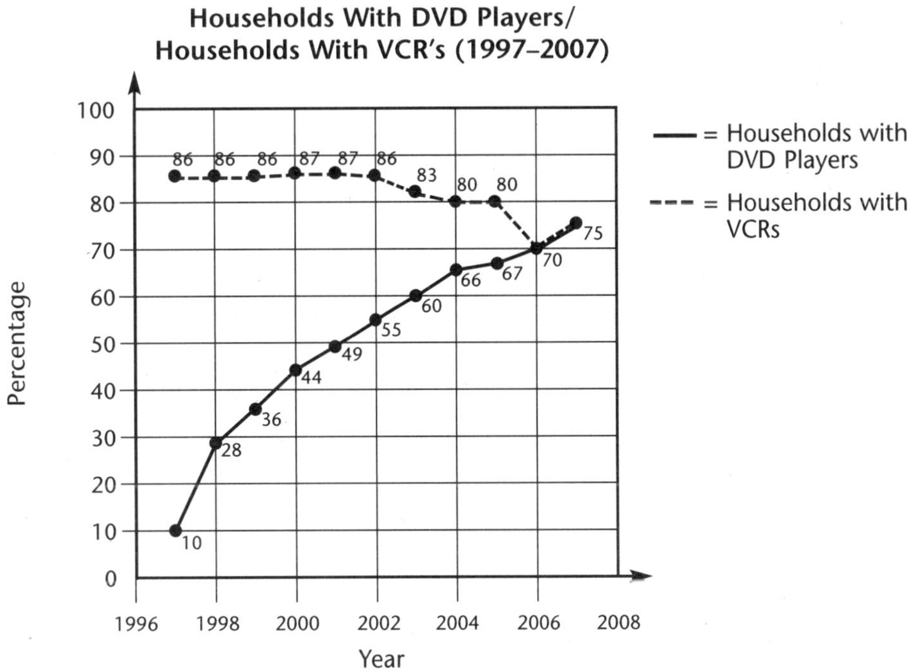

IV.C Solve problems by reading and interpreting collected data

1. For each survey, only 362,000 households were polled. How many households had DVD players in 1997?
   A 36,200
   B 101,360
   C 311,320
   D 325,800

2. If the predictions on the graph are correct, how many more households will have VCRs than DVD players in 2004?
   A 14
   B 50,680
   C 51,320
   D 289,600

3. During which year was there the greatest difference between the number of households that owned DVD players and VCRs?
   A 1997
   B 1999
   C 2001
   D 2006

4. Between which 2 years will there be no difference in the number of households with DVD players and VCRs?
   A 1999 and 2000
   B 2000 and 2001
   C 2003 and 2004
   D 2006 and 2007

©ECS Learning Systems, Inc.

All rights reserved

Notes

©ECS Learning Systems, Inc.

All rights reserved

Reasonableness

V. Evaluate the reasonableness of a solution to a problem.

A. Determine the reasonableness of a solution to a problem

©ECS Learning Systems, Inc.

All rights reserved

Objective 5: Pretest

V.A Determine the reasonableness of a solution to a problem (1–10)

1. Dr. Troy travels by plane to speak at conferences 7 times a year. He spends an average of 8.3 hours on the plane for each roundtrip he makes. Which is the most reasonable estimate for the amount of time Dr. Troy spends on planes in 1 year?

   A Fewer than 15 hours

   B Between 15 and 50 hours

   C Between 50 and 56 hours

   D More than 56 hours

2. When Seleta broke her arm, the hospital visit cost $459.34. Seleta's insurance company covered 85% of the cost. Which is the most reasonable estimate for the amount of money Seleta's insurance company paid for her hospital visit?

   A $450

   B $400

   C $325

   D $300

3. Mr. Skipper usually pays 12¢ per minute for each long-distance phone call he makes. In June, he received a discount and paid only 8¢ per minute for long-distance calls. If he spent 78 minutes making long-distance phone calls in June, which is a reasonable estimate for how much money Mr. Skipper saved with the discount?

   A $1

   B $2

   C $3

   D $4

4. Jenna drives on a toll road every day to get to work. Each month, she pays $35 for a 1-month pass that allows her to drive on the toll road at any time. People without the pass must pay $1.25 each way to travel to and from work daily. If Jenna works 22 days each month, which is the most reasonable estimate for how much money she saves by buying a pass?

   A More than $25

   B Between $20 and $25

   C Between $15 and $20

   D Less than $15

5. Duncan has only $5\frac{1}{2}$ hours between the time he gets home from school and his bedtime. If he spends an average of 1 hour eating dinner and $1\frac{2}{3}$ hours on homework and chores, a reasonable estimate for the amount of free time he has left is—

   A Less than 2 hours

   B Between 2 and $2\frac{1}{2}$ hours

   C Between $2\frac{1}{2}$ and 3 hours

   D More than 3 hours

©ECS Learning Systems, Inc.

All rights reserved

6. Mona purchased a movie ticket for \$7.25, popcorn for \$3.10, and a drink for \$2.95. She paid for all of her purchases with a \$20 bill. A reasonable estimate for the amount of change she received is—

   A \$6.25

   B \$7

   C \$7.75

   D \$8

7. A cheetah runs at a rate of 59.3 to 67.3 miles per hour for 37 minutes while chasing a herd of zebra. A reasonable estimate for the distance the cheetah ran is—

   A Less than 40 miles

   B Between 40 and 50 miles

   C Between 50 and 60 miles

   D More than 60 miles

8. In a medical study, doctors determined that most people dream 15% to 19% of the time when they sleep. If you sleep for 8 hours, a reasonable estimate for the amount of time you will spend dreaming is—

   A 1 hr

   B 1.3 hrs

   C 2.0 hrs

   D 2.3 hrs

9. Julian marinates 1 to 1.5 pounds of chicken in 5 ounces of oil and 3 teaspoons of spices. If he decides to marinate 8.7 pounds of chicken, what is a reasonable estimate for the amount of oil he will use?

   A Less than 30 ounces

   B Between 30 and 45 ounces

   C Between 45 and 60 ounces

   D More than 60 ounces

10. When Faith was sick, she took 2 tablespoons of medicine once every 4 to $5\frac{1}{2}$ hours. Which is a reasonable estimate for the total amount of medicine she took in 1 day (24 hours)?

    A Less than 6 tablespoons

    B Between 6 and 9 tablespoons

    C Between 9 and 12 tablespoons

    D More than 12 tablespoons

©ECS Learning Systems, Inc.

All rights reserved

Practice 5.A1

V.A Determine the reasonableness of a solution to a problem

1. June works from 7.9 to 8.4 hours each day, 6 days a week. She is paid for any time she works over 40 hours. A reasonable estimate for the amount of overtime June works weekly is—

   A Less than 7 hours

   B Between 7 and 8 hours

   C Between 8 and 9 hours

   D More than 9 hours

2. A caterer prepares for each person at a party to have 8 ounces of punch. If there are 213 guests at the party, a reasonable estimate for the amount of punch the caterer must make is—

   A 5 gal

   B 10 gal

   C 13 gal

   D 15 gal

3. Mrs. Diez earns a bonus worth 3% of her yearly income at the end of each year. If Mrs. Diez earns $32,985 in 1 year, which is the most reasonable estimate for the bonus she earns?

   A $90

   B $1,000

   C $3,000

   D $9,000

4. Keenan must replace the rope on a flagpole. He estimates that the flagpole is 19 to 22 feet tall. Keenan has 34 feet of rope. Which is the most reasonable estimate for how much more rope Keenan needs?

   A 6 ft

   B 9 ft

   C 12 ft

   D 15 ft

5. Portia usually scores 12 to 16 points in each basketball game she plays. There were 28 games in the season, but Portia was injured for 4 of the games. A reasonable estimate for the number of points she scored during the season is—

   A Less than 100 points

   B Between 100 and 200 points

   C Between 200 and 300 points

   D More than 300 points

©ECS Learning Systems, Inc.

All rights reserved

Practice 5.A2

V.A Determine the reasonableness of a solution to a problem

1. There are 28 teams from 12 countries entered in a European soccer tournament. If each soccer team has 11 to 13 players, what is a reasonable estimate for the total number of soccer players entered in the tournament?

- **A** 290
- **B** 350
- **C** 400
- **D** 450

2. At the zoo, Jeremy saw alligators that were 1.9 to 2.4 meters long. He also saw crocodiles that were 5.8 to 6.7 meters long. A reasonable estimate for the difference in the lengths of the alligators and crocodiles is—

- **A** Less than 4.5 meters
- **B** Between 4.5 and 5 meters
- **C** Between 5 and 5.5 meters
- **D** More than 5.5 meters

3. Cristina wants to replace the carpet in her $11\frac{1}{2}$-foot by $9\frac{3}{4}$-foot rectangular bedroom. What is a reasonable estimate for the area of carpet Cristina must buy to replace the old carpet in her room?

- **A** 120 ft^2
- **B** 300 ft^2
- **C** 900 ft^2
- **D** 1,200 ft^2

4. Mr. Derks spent 22 to 31 minutes grading each history exam. If 32 students took the exam, a reasonable estimate for the amount of time Mr. Derks spent grading all of the exams is—

- **A** Less than 10 hours
- **B** Between 10 and 13 hours
- **C** Between 13 and 16 hours
- **D** More than 16 hours

5. Matthew weighed 7 pounds 8 ounces when he was born. After 1 month, he weighed 9 pounds 3 ounces. If Matthew continued to gain the same amount of weight each month for 3 months, which is the most reasonable estimate for the total amount of weight he would have gained in 3 months?

- **A** 5 oz
- **B** 1 lb 10 oz
- **C** 3 lb
- **D** 5 lb

©ECS Learning Systems, Inc. All rights reserved

Practice 5.A3

V.A Determine the reasonableness of a solution to a problem

1. Felicity receives a $7 allowance each week. Last week, she spent $3.88 on makeup at the grocery store. She also spent $2.56 on nail polish. Which is the most reasonable estimate for the amount of money Felicity had left from last week's allowance?

   A 50¢

   B $1

   C $1.50

   D $2

2. Cassie had a dental appointment at 3:30 p.m. She arrived at the dentist's office at 3:15 p.m. and did not leave for another 2 hours 52 minutes. Which is the most reasonable estimate for what time Cassie left the dentist's office?

   A 4:30 p.m.

   B 5:00 p.m.

   C 5:45 p.m.

   D 6:15 p.m.

3. Every day, Drew exercises for 37 to 45 minutes. A reasonable estimate for the amount of time Drew exercises during an entire month is—

   A Less than 15 hours

   B Between 15 and 18 hours

   C Between 18 and 21 hours

   D More than 21 hours

4. Mr. Dryer wants to make ice cream sundaes at his son's birthday party. Each child will receive 1 to $1\frac{1}{2}$ cups of ice cream. If there are 16 children at the party, which is a reasonable estimate for the amount of ice cream Mr. Dryer should purchase?

   A Less than $\frac{1}{2}$ gallon

   B Between $\frac{1}{2}$ and 1 gallon

   C Between 1 and $1\frac{1}{2}$ gallons

   D More than $1\frac{1}{2}$ gallons

5. A real estate agent wants to sell a house for anywhere from $154,800 to $165,350. She will earn a 3% commission of the total sale of the house. A reasonable estimate for the commission the real estate agent can earn on the sale of the house is—

   A Less than $4,500

   B Between $4,500 and $4,900

   C Between $4,900 and $5,300

   D More than $5,300

©ECS Learning Systems, Inc.

All rights reserved

Practice 5.A4

V.A Determine the reasonableness of a solution to a problem

1. At a doughnut shop, 1 doughnut costs 60¢ to 65¢. A dozen doughnuts costs $4.79. A reasonable estimate for the difference in the price of a doughnut purchased by itself or in a package of a dozen is—

   A Less than 10¢

   B Between 10¢ and 15¢

   C Between 15¢ and 20¢

   D More than 20¢

2. Sundara saves $3.85 each week in order to purchase a pair of shoes that costs $29.88. What is a reasonable estimate for how many weeks Sundara must save in order to purchase the shoes?

   A Less than 6

   B Between 6 and 7

   C Between 7 and 8

   D More than 8

3. Mitch travels 13.1 miles to work each day. He takes a different route home that is 14.5 miles long. If Mitch works 5 days a week, a reasonable estimate for the number of miles he drives to and from work each week is—

   A 100

   B 120

   C 140

   D 150

4. Railroad track weighed 65 to 67 kilograms per meter in the 1930s. In the 1990s, track weighed 74 to 76 kilograms per meter. A reasonable estimate for the difference in the weight of track laid in 1930 and 1990 over a 28-meter distance is—

   A 150 kg

   B 200 kg

   C 280 kg

   D 350 kg

5. Chauncey can read a 625- to 767-page book in 2 weeks. Which is a reasonable estimate for the number of pages Chauncey can read each day?

   A 60

   B 50

   C 40

   D 30

6. Daphne sold her used car for 86% of her asking price of $5,250. A reasonable estimate for the amount of money Daphne sold her car for is—

   A $4,500

   B $4,900

   C $5,100

   D $5,200

©ECS Learning Systems, Inc.

All rights reserved

Practice 5.A5

V.A Determine the reasonableness of a solution to a problem

1. Dylan went to the theater to watch a movie. The movie was 143 minutes long, and the previews before the movie were 28 minutes long. A reasonable estimate for the total amount of time Dylan was in the theater is—

   A 2 hr

   B $2\frac{1}{2}$ hr

   C 3 hr

   D $3\frac{1}{2}$ hr

2. Curtis' dog, Duchess, eats 4.4 to 4.6 ounces of dog food each day. If 1 bag of dog food contains 12 pounds of food, a reasonable estimate for the number of days the bag of dog food will last is—

   A Less than 30

   B Between 30 and 35

   C Between 35 and 40

   D More than 40

3. Deputy Fife drives a patrol route that is 75.9 miles long. If he goes on patrol 18 times a month, a reasonable estimate for the number of miles he travels on patrol in 1 year is—

   A 1,500 mi

   B 15,000 mi

   C 16,000 mi

   D 17,600 mi

4. Perfect Print Shop prints color magazines at a rate of 8 cents per page. If a publishing company hired them to print a 48-page magazine and received a 10% discount, a reasonable estimate for the amount of money the publishing company would pay for every 10 copies of the magazine is—

   A $36

   B $40

   C $50

   D $52

5. Suki wanted to paint the 4 exterior walls of her storage shed. The shed is 3.1 meters long, 4.2 meters tall, and 2.67 meters wide. If 1 pint of paint covers 0.8 to 0.85 square meters of wall space, a reasonable estimate for the amount of paint Suki will need is—

   A Less than 8 gallons

   B Between 8 and 9 gallons

   C Between 9 and 10 gallons

   D More than 10 gallons

©ECS Learning Systems, Inc.

All rights reserved

Practice 5.A6

V.A Determine the reasonableness of a solution to a problem

1. Susan wanted to purchase a new outfit for the first day of school. She saved $42.87 over the summer and then bought a shirt for $21.96 and a skirt for $18.75. A reasonable estimate for the amount of money Susan had left after buying the outfit is—

   A $20

   B $18

   C $2

   D $1

2. A.J. must have 124 hours of class credit to earn his college degree. If he wants to finish college in 8 semesters, a reasonable estimate for the number of hours he must pass each semester is—

   A Less than 10

   B Between 10 and 12

   C Between 12 and 15

   D More than 15

3. Michelle practices running the 100-meter dash at least 7 times each day, 6 days a week. A reasonable estimate for the total distance Michelle runs in 1 month (4 weeks) practicing the 100-meter dash is—

   A Less than 16 kilometers

   B Between 16 and 17 kilometers

   C Between 17 and 18 kilometers

   D More than 18 kilometers

4. Dennis was thirsty after he built a sandcastle at the beach. What is the best estimate for how much water Dennis would drink at one time to quench his thirst?

   A 56 mL

   B 123 mL

   C 0.6 L

   D 4 L

5. Mrs. Carter wants to put new floor-length curtains along each side of her 7 living room windows. Each window is 10 feet high. Which is a reasonable estimate for how many yards of fabric she should buy?

   A Less than 20 yards

   B Between 20 and 50 yards

   C Between 50 and 70 yards

   D More than 70 yards

6. The new band, In-Tune, wants to record a CD with 9 songs on it. A reasonable estimate for the time it would take to listen to the entire CD is—

   A 15 min

   B 35 min

   C 1 hr 15 min

   D 2 hr 15 min

©ECS Learning Systems, Inc. All rights reserved

Notes

©ECS Learning Systems, Inc.

All rights reserved

Appendix

- Answer Keys
- Teaching Tips
- Answer Sheets

©ECS Learning Systems, Inc.

All rights reserved

Answer Key: Using Operations to Solve Problems

Objective 1 Pretest (p. 12)

1. C	2. B	3. D	4. D	5. A
6. B	7. D	8. C	9. B	10. A
11. C	12. D	13. B	14. C	15. D
16. A	17. B	18. C	19. D	20. D
21. B	22. A	23. A	24. C	

Practice 1.A1 (p. 16)

1. B 2. A 3. C 4. C 5. B

Practice 1.A2 (p. 17)

1. A 2. C 3. B 4. B 5. D

Practice 1.A3 (p. 18)

1. C 2. C 3. D 4. B 5. B
6. D

Practice 1.A4 (p. 19)

1. D 2. A 3. D 4. C 5. B

Practice 1.A5 (p. 20)

1. C 2. D 3. D 4. B 5. B

Practice 1.A6 (p. 21)

1. C 2. A 3. C 4. B 5. D
6. D

Practice 1.A7 (p. 22)

1. D 2. B 3. A 4. C 5. A
6. C

Practice 1.B1 (p. 23)

1. B 2. C 3. D 4. A 5. D
6. C

Practice 1.B2 (p. 24)

1. C 2. B 3. B 4. D 5. A
6. B

Practice 1.B3 (p. 25)

1. C 2. A 3. B 4. C 5. D
6. A

Practice 1.B4 (p. 26)

1. C 2. B 3. D 4. B 5. C
6. A

Practice 1.B5 (p. 27)

1. C 2. B 3. C 4. D 5. A
6. B

Practice 1.B6 (p. 28)

1. C 2. B 3. D 4. C 5. B
6. C

Practice 1.B7 (p. 29)

1. C 2. D 3. D 4. A 5. D
6. C

©ECS Learning Systems, Inc.

All rights reserved

Answer Key: Estimating Solutions

Objective 2 Pretest (p. 32)

1. B	2. C	3. D	4. A	5. C
6. D	7. B	8. C	9. D	10. C
11. C	12. B	13. B	14. D	15. B
16. C	17. A	18. C	19. A	20. B
21. D	22. B	23. C	24. C	25. B
26. B	27. D	28. D	29. A	30. C

Practice 2.A1 (p. 38)

1. C 2. B 3. A 4. D 5. C
6. A 7. A

Practice 2.A2 (p. 39)

1. D 2. B 3. C 4. B 5. D
6. C

Practice 2.A3 (p. 40)

1. B 2. C 3. B 4. B 5. C

Practice 2.A4 (p. 41)

1. A 2. D 3. C 4. C 5. A
6. B

Practice 2.A5 (p. 42)

1. B 2. C 3. B 4. C 5. B
6. A

Practice 2.B1 (p. 43)

1. A 2. B 3. C 4. A 5. D
6. B

Practice 2.B2 (p. 44)

1. C 2. B 3. B 4. C 5. D
6. C

Practice 2.B3 (p. 45)

1. C 2. A 3. B 4. B 5. A
6. D

Practice 2.B4 (p. 46)

1. B 2. B 3. C 4. D 5. C
6. A

Practice 2.B5 (p. 47)

1. B 2. D 3. C 4. B 5. C
6. B

Practice 2.B6 (p. 48)

1. D 2. B 3. C 4. B 5. B
6. A

Practice 2.C1 (p. 49)

1. C 2. B 3. C 4. A 5. C
6. C

Practice 2.C2 (p. 50)

1. B 2. C 3. D 4. A 5. C
6. C

Practice 2.C3 (p. 51)

1. D 2. A 3. A 4. C 5. B
6. A

Practice 2.C4 (p. 52)

1. C 2. C 3. A 4. D 5. C
6. B

Practice 2.C5 (p. 53)

1. C 2. A 3. C 4. D 5. C

Practice 2.C6 (p. 54)

1. D 2. A 3. C 4. B 5. D
6. A

©ECS Learning Systems, Inc.

All rights reserved

Answer Key: Solution Strategies/ Analyze or Solve Problems

Objective 3 Pretest (p. 57)

1. B 2. C 3. D 4. A 5. C
6. D 7. B 8. C 9. A 10. A
11. C 12. D 13. C 14. B 15. B
16. C 17. A 18. C 19. D 20. B
21. B 22. D 23. C 24. B

Practice 3.A1 (p. 61)

1. A 2. D 3. D 4. D 5. B
6. B

Practice 3.A2 (p. 62)

1. B 2. C 3. C 4. A 5. B
6. A

Practice 3.A3 (p. 63)

1. C 2. B 3. C 4. D 5. C
6. A

Practice 3.A4 (p. 64)

1. C 2. B 3. C 4. A 5. A
6. C

Practice 3.A5 (p. 65)

1. D 2. B 3. C 4. D 5. B
6. C

Practice 3.B1 (p. 66)

1. D 2. B 3. B 4. A 5. B
6. C

Practice 3.B2 (p. 67)

1. D 2. B 3. D 4. B 5. B
6. C

Practice 3.B3 (p. 68)

1. B 2. C 3. C 4. D 5. B
6. C

Practice 3.B4 (p. 69)

1. B 2. B 3. C 4. D 5. C
6. D

Practice 3.B5 (p. 70)

1. B 2. A 3. C 4. C 5. B
6. D

Practice 3.C1 (p. 71)

1. D 2. C 3. D 4. C

Practice 3.C2 (p. 72)

1. C 2. D 3. B 4. D

Practice 3.C3 (p. 73)

1. C 2. B 3. B 4. A

Practice 3.C4 (p. 74)

1. A 2. C 3. D 4. B

Practice 3.C5 (p. 75)

1. A 2. C 3. A 4. A

©ECS Learning Systems, Inc.

All rights reserved

Answer Key: Mathematical Representation

Objective 4 Pretest (p. 78)

1. B	2. A	3. D	4. B	5. C
6. C	7. A	8. D	9. B	10. C
11. C	12. D	13. D	14. A	15. B
16. C	17. D			

Practice 4.A1 (p. 82)
1. B 2. C 3. C 4. A

Practice 4.A2 (p. 83)
1. C 2. B 3. D 4. A

Practice 4.A3 (p. 84)
1. C 2. C 3. B 4. A

Practice 4.A4 (p. 85)
1. C 2. B 3. A 4. D

Practice 4.A5 (p. 86)
1. C 2. B 3. D 4. A

Practice 4.A6 (p. 87)
1. B 2. C 3. B 4. D

Practice 4.B1 (p. 88)
1. C 2. D 3. A 4. C

Practice 4.B2 (p. 89)
1. D 2. C 3. A 4. C

Practice 4.B3 (p. 90)
1. B 2. A 3. D 4. A

Practice 4.B4 (p. 91)
1. D 2. D 3. A 4. A

Practice 4.B5 (p. 92)
1. C 2. D 3. A 4. D

Practice 4.B6 (p. 93)
1. A 2. D 3. C 4. B

Practice 4.B7 (p. 94)
1. D 2. A 3. B 4. B

Practice 4.B8 (p. 95)
1. C 2. B 3. A 4. C

Practice 4.C1 (p. 96)
1. A 2. B 3. B 4. C

Practice 4.C2 (p. 97)
1. D 2. B 3. B 4. A

Practice 4.C3 (p. 98)
1. A 2. B 3. C 4. C

Practice 4.C4 (p. 99)
1. A 2. C 3. C 4. B

Practice 4.C5 (p. 100)
1. D 2. A 3. B 4. B

Practice 4.C6 (p. 101)
1. A 2. D 3. D 4. C

Practice 4.C7 (p. 102)
1. C 2. D 3. A 4. A

Practice 4.C8 (p. 103)
1. A 2. B 3. A 4. D

©ECS Learning Systems, Inc. All rights reserved

Answer Key: Reasonableness

Objective 5 Pretest (p. 106)

1. D	2. B	3. C	4. C	5. C
6. B	7. A	8. B	9. B	10. C

Practice 5.A1 (p. 108)

1. C	2. C	3. B	4. B	5. D

Practice 5.A2 (p. 109)

1. B	2. A	3. A	4. C	5. D

Practice 5.A3 (p. 110)

1. A	2. D	3. C	4. C	5. B

Practice 5.A4 (p. 111)

1. D	2. C	3. C	4. C	5. B
6. A				

Practice 5.A5 (p. 112)

1. C	2. D	3. D	4. A	5. A

Practice 5.A6 (p. 113)

1. C	2. D	3. B	4. C	5. B
6. C				

©ECS Learning Systems, Inc.

All rights reserved

A Shadow on the Walk

Objective: Analyze or solve problems involving similarity and the Pythagorean Theorem

Materials: marking pens, recording sheets, large sheets of butcher paper, several objects that can be measured while standing on the ground (e.g., lamp, plant, thermos, trophy), measuring tape, calculators

Directions: On a sunny day, place large sheets of butcher paper on a sidewalk or level ground. Have students work in groups of 2 or 3. Each group chooses 2 objects (e.g., thermos and trophy) to measure. The group measures and records the height of only one object. Groups then place both objects side by side on the butcher paper so that each object casts a shadow. The group measures and records the length of each shadow. Using the 3 measurements they have taken, each group determines the height of the object they did not measure.

To solve the problem, students should set up a ratio like the one below.

$$\frac{\text{Height of object 1}}{\text{Length of shadow 1}} = \frac{\text{Height of object 2}}{\text{Length of shadow 2}}$$

Vary the activity by having students measure both objects, but measure the shadow of only one. The students must then determine the length of the shadow that would be cast by the other object.

©ECS Learning Systems, Inc.

All rights reserved

Chart Buster

Objective: Use the operations of multiplication and division to solve problems involving fractions and decimals

Materials: pencil, paper, operation charts (one per student)

Directions: Prepare an operations chart like the one shown below.

	2.8	3.6	4.05	6.15
+1.4				
–2.1				
x0.04				
÷0.6				

Give one chart to each student. Students complete the chart by performing each operation indicated and recording the answer on the chart. Below you will see the first column of answers for the chart introduced above.

	2.8	3.6	4.05	6.15
+1.4	4.2			
–2.1	0.7			
x0.04	0.112			
÷0.6	$4.\overline{6}$			

Students should work as quickly as possible. Mental calculations are allowed and recommended. Check answers with students when they have finished.

Vary the activity by providing different decimals and fractions on an operations chart. A chart may have decimals only, fractions only, or a mixture of both. Students may also create their own charts and challenge their classmates to solve them.

©ECS Learning Systems, Inc. All rights reserved

Mental Math Minutes

Objective: Use the operations of multiplication and division to estimate answers to problems

Materials: pencil, paper, sample problems

Directions: In this activity, students develop and practice estimating answers before actually solving problems.

Display a problem for students to read.

Example:

Last summer, Morgan earned money by baby-sitting. She earned an average of $23.70 each week. If she worked a total of 9 weeks, how much money did Morgan earn by baby-sitting?

Have the students estimate the answer to the problem and use the following sentence stem to give their responses:

The correct answer is less than __________, but greater than __________.

Students should give the most accurate range as possible, without actually solving the problem.

Sample answer:

The correct answer is less than $240, but greater than $220.

After students have written their answers, discuss their responses. Have students explain how they determined their estimated answers.

Repeat the process with several other problems.

Extension: Have students find exact answers to each problem and compare these to their estimations.

©ECS Learning Systems, Inc.

All rights reserved

Eliminate the Unnecessary

Objective: Formulate strategies to solve problems involving decimals and fractions

Materials: overhead projector, overhead pens (blue, green, red), sample problems with both extraneous information and missing information needed to solve the problem

Directions: Divide the class into two teams. Give each team a red marker. Then give one team a blue marker and the other team a green marker. Have each team name a captain who will mark the team's answers. Project a problem on the overhead projector.

Team members work together to determine the extraneous information in the problem. Taking turns, each team has the opportunity to cross out information that is not pertinent to solving the problem, using either the green or blue marker.

Team members must also determine missing information needed to solve the problem. The first team to write (in red) the additional information needed to solve the problem earns one point.

Sample problem:

A group of Boy Scouts is going camping. One adult leader must buy flashlights for each camper and adult leader going on the trip. Two scouts will sleep in each tent. The 4 adult leaders supervising the camping trip will sleep in a separate tent. The weather will be cool, so each scout will need 2 pairs of socks and a sweater. What other information is necessary to determine the total number of flashlights the adult leader should buy for the camping trip?

Sample Solution

A group of Boy Scouts is going camping. One adult leader must buy flashlights for each camper and adult leader going on the trip. There will be 4 adult leaders supervising the camping trip. What other information is necessary to determine the total number of flashlights the adult leader should buy for the camping trip?

Added information: The number of Boy Scouts going on the camping trip.

Repeat the process with several other problems, increasing the level of difficulty from problem to problem.

©ECS Learning Systems, Inc.

All rights reserved

Math Web Sites

The following list of Web sites provides a variety of math information and learning oppportunities. Although some Web sites come and go at a rapid pace, the sites listed in the book were "live" at the time of publication.

National Council of Teachers of Mathematics
http://www.nctm.org

Homepage for New Math Teachers
http://people.clarityconnect.com/webpages/terri/terri.html

K-12 Teaching Materials
http://archives.math.utk.edu/k12.html

Platonic Realms
http://www.mathacademy.com

Texas Instrument Calculator Home Page
http://education.ti.com/

Ask Dr. Math!
http://mathforum.org/dr.math/

Cool Math
http://www.coolmath.com/

MathMagic on the Web!
http://mathforum.org/mathmagic/

Explorer Page for Mathematics
http://unite.ukans.edu/

Mathematics Lesson Plans
http://math.rice.edu/~lanius/Lessons/

Mathematics, Science, and Technology Projects and Activities
http://www.mste.uiuc.edu/

MATHCOUNTS Home Page
http://www.mathcounts.org/

Mathematical Quotation Server
http://math.furman.edu/~mwoodard/mquot.html

Geometry Junkyard
http://www.ics.uci.edu/~eppstein/junkyard

K-12 Internet Geometry Resources
http://mathforum.org/geometry/k12.geometry.html

©ECS Learning Systems, Inc.

All rights reserved

Formula Chart

Perimeter	square	$P = 4s$
	rectangle	$P = 2(l + w)$
Circumference	circle	$C = 2\pi r$
Area	square	$A = s^2$
	rectangle	$A = lw$ or $A = bh$
	triangle	$A = \frac{bh}{2}$
	trapezoid	$A = \frac{1}{2}(b_1 + b_2)h$
	circle	$A = \pi r^2$
Surface area	cube	$S = 6s^2$
	cylinder (lateral)	$S = 2\pi rh$
Volume	rectangular prism	$V = lwh$
	cylinder	$V = \pi r^2 h$
	cube	$V = s^3$
Pythagorean Theorem	right triangle	$a^2 + b^2 = c^2$

Measurement Conversions

	Metric	**Customary**
Length	1 kilometer = 1,000 meters 1 meter = 100 centimeters 1 centimeter = 10 millimeters	1 mile =1,760 yards 1 mile = 5,280 feet 1 yard = 3 feet 1 foot = 12 inches
Volume and Capacity	1 liter = 1,000 milliliters	1 gallon = 4 quarts 1 gallon = 128 ounces 1 quart = 2 pints 1 pint = 2 cups 1 cup = 8 ounces
Weight and Mass	1 kilogram = 1,000 grams 1 gram = 1,000 milligrams	1 ton = 2,000 pounds 1 pound = 16 ounces
Time		1 year = 12 months 1 year = 52 weeks 1 year = 365 days 1 week = 7 days 1 day = 24 hours 1 hour = 60 minutes 1 minute = 60 seconds

©ECS Learning Systems, Inc. All rights reserved

Name ______________________________ Date ___________

Objective # ________ Pretest

Pretest Answer Sheet

1. Ⓐ Ⓑ Ⓒ Ⓓ	11. Ⓐ Ⓑ Ⓒ Ⓓ	21. Ⓐ Ⓑ Ⓒ Ⓓ	31. Ⓐ Ⓑ Ⓒ Ⓓ
2. Ⓐ Ⓑ Ⓒ Ⓓ	12. Ⓐ Ⓑ Ⓒ Ⓓ	22. Ⓐ Ⓑ Ⓒ Ⓓ	32. Ⓐ Ⓑ Ⓒ Ⓓ
3. Ⓐ Ⓑ Ⓒ Ⓓ	13. Ⓐ Ⓑ Ⓒ Ⓓ	23. Ⓐ Ⓑ Ⓒ Ⓓ	33. Ⓐ Ⓑ Ⓒ Ⓓ
4. Ⓐ Ⓑ Ⓒ Ⓓ	14. Ⓐ Ⓑ Ⓒ Ⓓ	24. Ⓐ Ⓑ Ⓒ Ⓓ	34. Ⓐ Ⓑ Ⓒ Ⓓ
5. Ⓐ Ⓑ Ⓒ Ⓓ	15. Ⓐ Ⓑ Ⓒ Ⓓ	25. Ⓐ Ⓑ Ⓒ Ⓓ	35. Ⓐ Ⓑ Ⓒ Ⓓ
6. Ⓐ Ⓑ Ⓒ Ⓓ	16. Ⓐ Ⓑ Ⓒ Ⓓ	26. Ⓐ Ⓑ Ⓒ Ⓓ	36. Ⓐ Ⓑ Ⓒ Ⓓ
7. Ⓐ Ⓑ Ⓒ Ⓓ	17. Ⓐ Ⓑ Ⓒ Ⓓ	27. Ⓐ Ⓑ Ⓒ Ⓓ	37. Ⓐ Ⓑ Ⓒ Ⓓ
8. Ⓐ Ⓑ Ⓒ Ⓓ	18. Ⓐ Ⓑ Ⓒ Ⓓ	28. Ⓐ Ⓑ Ⓒ Ⓓ	38. Ⓐ Ⓑ Ⓒ Ⓓ
9. Ⓐ Ⓑ Ⓒ Ⓓ	19. Ⓐ Ⓑ Ⓒ Ⓓ	29. Ⓐ Ⓑ Ⓒ Ⓓ	39. Ⓐ Ⓑ Ⓒ Ⓓ
10. Ⓐ Ⓑ Ⓒ Ⓓ	20. Ⓐ Ⓑ Ⓒ Ⓓ	30. Ⓐ Ⓑ Ⓒ Ⓓ	40. Ⓐ Ⓑ Ⓒ Ⓓ

©ECS Learning Systems, Inc. All rights reserved

Practice Answer Sheet **Name** ______________________________

Date: ________ Practice: # __________	Date: ________ Practice: # __________
1. Ⓐ Ⓑ Ⓒ Ⓓ	1. Ⓐ Ⓑ Ⓒ Ⓓ
2. Ⓐ Ⓑ Ⓒ Ⓓ	2. Ⓐ Ⓑ Ⓒ Ⓓ
3. Ⓐ Ⓑ Ⓒ Ⓓ	3. Ⓐ Ⓑ Ⓒ Ⓓ
4. Ⓐ Ⓑ Ⓒ Ⓓ	4. Ⓐ Ⓑ Ⓒ Ⓓ
5. Ⓐ Ⓑ Ⓒ Ⓓ	5. Ⓐ Ⓑ Ⓒ Ⓓ
6. Ⓐ Ⓑ Ⓒ Ⓓ	6. Ⓐ Ⓑ Ⓒ Ⓓ
7. Ⓐ Ⓑ Ⓒ Ⓓ	7. Ⓐ Ⓑ Ⓒ Ⓓ
8. Ⓐ Ⓑ Ⓒ Ⓓ	8. Ⓐ Ⓑ Ⓒ Ⓓ
Date: ________ Practice: # __________	Date: ________ Practice: # __________
1. Ⓐ Ⓑ Ⓒ Ⓓ	1. Ⓐ Ⓑ Ⓒ Ⓓ
2. Ⓐ Ⓑ Ⓒ Ⓓ	2. Ⓐ Ⓑ Ⓒ Ⓓ
3. Ⓐ Ⓑ Ⓒ Ⓓ	3. Ⓐ Ⓑ Ⓒ Ⓓ
4. Ⓐ Ⓑ Ⓒ Ⓓ	4. Ⓐ Ⓑ Ⓒ Ⓓ
5. Ⓐ Ⓑ Ⓒ Ⓓ	5. Ⓐ Ⓑ Ⓒ Ⓓ
6. Ⓐ Ⓑ Ⓒ Ⓓ	6. Ⓐ Ⓑ Ⓒ Ⓓ
7. Ⓐ Ⓑ Ⓒ Ⓓ	7. Ⓐ Ⓑ Ⓒ Ⓓ
8. Ⓐ Ⓑ Ⓒ Ⓓ	8. Ⓐ Ⓑ Ⓒ Ⓓ

©ECS Learning Systems, Inc. All rights reserved